Lecture Notes in Mathematics

Edited by A. Dold and B. Eckmann

698

Emil Grosswald

Bessel Polynomials

Springer-Verlag
Berlin Heidelberg New York 1978

Author
Emil Grosswald
Department of Mathematics
Temple University
Philadelphia, PA 19122/USA

AMS Subject Classifications (1970): primary: 33 A 70
secondary: 33 A 65, 33 A 40, 33 A 75, 33 A 45, 33-01, 33-02, 33-03, 35 J 05,
41 A 10, 44 A 10, 30 A 22, 30 A 80, 30 A 84, 10 F 35, 12 A 20, 12 D 10, 60 E 05

ISBN 3-540-09104-1 Springer-Verlag Berlin Heidelberg New York
ISBN 0-387-09104-1 Springer-Verlag New York Heidelberg Berlin

© by Springer-Verlag Berlin Heidelberg 1978
Printed in Germany

Printing and binding: Beltz Offsetdruck, Hemsbach/Bergstr.
2141/3140-543210

To

ELIZABETH

BLANCHE

and

VIVIAN

FOREWORD

The present book consists of an Introduction, 15 Chapters, an Appendix, two Bibliographies and two Indexes.The chapters are numbered consecutively, from 1 to 15 and are grouped into four parts, as follows:

Part I - A short historic sketch (1 Chapter) followed by the basic theory (3 Chapters);

Part II - Analytic properties (5 Chapters);

Part III - Algebraic properties (4 Chapters);

Part IV - Applications and miscellanea (2 Chapters).

According to its subject matter, the chapter on asymptotic properties would fit better into Part II; however, some of the proofs require results obtained only in Chapter 10 (properties of zeros) and, for that reason, the chapter has been incorporated into Part III.

The Appendix contains a list of some 12 open problems.

In the first bibliography are listed all papers, monographs, etc., that could be located and that discuss Bessel Polynomials. It is quite likely that, despite all efforts made, absolute completeness has not been achieved. The present writer takes this opportunity to apologize to all authors, whose work has been overlooked.

A second, separate bibliography lists books and papers quoted in the text, but not directly related to Bessel Polynomials.

References to the bibliographies are enclosed in square brackets. Those refering to the second bibliography are distinguished by heavy print. So [1] refers to: W.H. Abdi - A basic analog of the Bessel Polynomials; while [**1**] refers to: M. Abramowitz and I.E. Segun - Handbook of Mathematical Functions.

Within each chapter, the sections, theorems, lemmata, corollaries, drawings, and formulae are numbered consecutively. If quoted, or referred to within the same chapter, only their own number is mentioned. If, e.g., in Chapter 10 a reference is made to formula (12), or to Section 2, this means formula (12), or Section 2 of Chapter 10. The same formula, or section quoted in another chapter, would be referred to as formula (10.12), or Section (10.2), respectively. The same holds, mutatis mutandis, for theorems, drawings, etc.

While writing this book, the author has received invaluable help from many colleagues; to all of them he owes a great debt of gratitude. Of particular importance was the great moral support received from Professors H.L. Krall and O. Frink, as well as A.M. Krall. Professors Krall also read most of the manuscript and made valuable suggestions for improvements.

As already mentioned, there is no hope for an absolutely complete bibliography; however, many more omissions would have occurred, were it not for the help received, in addition to the mentioned colleagues, also from Professors R.P. Agarwal, W.A. Al-Salam, H.W. Gould, M.E.H. Ismail, C. Underhill, and A. Wragg.

Last, but not least, thanks are due to Ms. Gerry Sizemore-Ballard, for her skill and infinite patience in typing the manuscript and to my daughter Vivian for her help with the Indexes.

Part of the work on this book was done during the summer 1976, under a Summer Research Grant offered by Temple University and for which the author herewith expresses his gratitude.

July 1978 E. Grosswald

TABLE OF CONTENTS

INTRODUCTION

Let us look at a few problems that, at first view, have little in common.

PROBLEM 1: To prove that if $r = a/b$ is rational, then e^r is irrational; also that π is irrational.

Following C.L. Siegel **[53]** (who streamlined an idea due to Hermite), one first determines two polynomials $A_n(x)$ and $B_n(x)$, both of degree n, such that $e^x + A_n(x)/B_n(x)$ has a zero of order (at least) $2n + 1$ at $x = 0$. This means, in particular, that the power series expansion of $R_n(x) = B_n(x)e^x + A_n(x)$ starts with the term of degree $2n + 1$, $R_n(x) = c_1x^{2n+1} + c_2x^{2n+2} + \ldots$ say. By counting the number of conditions and the number of available coefficients, it turns out that $A_n(x)$ and $B_n(x)$ are uniquely defined, up to a multiplicative constant. By proper choice of this constant one can obtain that $A_n(x)$ and $B_n(x)$ should have integer coefficients. By simple manipulations one shows that $A_n(-x) = -B_n(x)$ and that $R_n(x) = (n!)^{-1}x^{2n+1} \int_0^1 t^n(1-t)^n e^{tx}dt$. The last assertions are proved by effective construction of the polynomials involved (see Sections 14.2 and 14.3 for details). It follows from the integral representation that $|R_n(x)| \leq (n!)^{-1}|x|^{2n+1}e^{|x|}$ and that $R_n(x) > 0$ for $x \geq 0$. If now $e^r = e^{a/b}$ were rational, also e^a would be rational; let $q > 0$ be its denominator. As already observed, $B_n(a)$ and $A_n(a)$ are integers, so that

$$m = qR_n(a) = q(B_n(a)e^a + A_n(a)) \text{ is a positive integer.}$$

Using the bound on $R_n(a)$, $0 < m < q.(n!)^{-1}a^{2n+1}e^a$ and, by Stirling's formula, $0 < m < q(a^{2n+1}e^a/n^{n+1/2}e^{-n}(2\pi)^{1/2})(1+\varepsilon)$, where $\varepsilon \to 0$ as $n \to \infty$. For sufficiently large n, $0 < m < 1$, which is absurd, because m is an integer. Hence, e^r cannot be rational.

Next setting $x = \pi i$,

$$R_n(\pi i) = -A_n(-\pi i)(-1) + A_n(\pi i) = (-1)^{n+1}\frac{\pi^{2n+1}}{n!}\int_0^1 t^n(1-t)^n \sin \pi t \, dt$$

(the last equality depends on some computations and will be justified in Chapter 14). The integrand is positive, so that $R_n(\pi i) \neq 0$. Let $k = [\frac{n}{2}]$ where $[x]$ stands for the greatest integer function; then $A_n(x) + A_n(-x)$ is a polynomial in x^2 of degree k and with integer coefficients. Hence, if π^2 is rational, with denominator $q > 0$, then $q^kR_n(\pi i) = q^k\{A_n(\pi i) + A_n(-\pi i)\} = m$, an integer, possibly negative, but certainly

$\neq 0$. Also, by using the integral representation of $R_n(x)$, $0 < |m| = q^k |R_n(\pi i)| \leq (n!)^{-1} q^k \pi^{2n+1} < \frac{(q^{1/2}\pi^2)^n \pi}{n^{n+1/2} e^{-n} (2\pi)^{1/2}} (1+\varepsilon)$ $(\varepsilon \to 0 \text{ for } n \to \infty)$, or $0 < |m| < 1$ for sufficiently large n. This is, of course impossible for integral m. Hence π^2, and a fortiori π are irrational. A (highly nontrivial) modification of this proof permits one to show much more, namely that e^r is actually transcendental for real, rational r. In particular, for r = 1, this implies the transcendency of e itself.

PROBLEM 2. To prove that the Student t-distribution of 2n+1 degrees of freedom is infinitely divisible.

We do not have to enter here into the probabilistic relevance, or even into the exact meaning of this important problem. Suffice it to say that, based on the pioneering work of Paul Levy [**44**] and of Gnedenko and Kolmogorov [**19**], Kelker [65] and then Ismail and Kelker [60] proved that the property holds if, and only if the function $\varphi(x) = \frac{K_{n-1/2}(\sqrt{x})}{\sqrt{x}\, K_{n+1/2}(\sqrt{x})}$ is completely monotonic on $[0,\infty)$, which means that $(-1)^k \varphi^{(k)}(x) \geq 0$ for $0 < x < \infty$ and all integral $k \geq 0$. Now, it is well-known (see, e.g. [**1**], 10.2.17) that if the index of $K_\nu(z)$ (the so called modified Hankel function) is of the form n+1/2 (n an integer), then $(2z/\pi)^{1/2} e^z K_{n+1/2}(z) = p_n(1/z)$, with $p_n(u)$ a polynomial of exact degree n. Previous relation can now be written as $\varphi(x) = \frac{p_{n-1}(x^{-1/2})}{x^{1/2} p_n(x^{-1/2})}$, or, with $P_n(u) = u^n p_n(1/u)$, $\varphi(x) = \frac{P_{n-1}(x^{1/2})}{P_n(x^{1/2})}$. We now use Bernstein's theorem (see [**68**]); this asserts that $\phi(x)$ is completely monotonic if, and only if it is the Laplace transform of a function G(t), non-negative on $0 < t < \infty$. In the present case it is possible to study the polynomials $P_n(x)$ and compute G(t). It turns out that $G(t) \geq 0$ for small $t \geq 0$ and also for t sufficiently large. This alone is not quite sufficient to settle the problem, but if we also knew that G(t) is monotonic, then the conclusion immediately follows. In fact, by playing around with G(t), one soon suspects that it is not only monotonic, but actually completely monotonic. In order to prove this, one appeals once more to Bernstein's theorem and finds that G(t) is the Laplace transform of $\phi(x) = (\pi^2 x)^{-1/2}\{1 + \sum_{j=1}^{n} \alpha_j (x + \alpha_j^2)^{-1}\}$, where $\alpha_1, \alpha_2, \ldots, \alpha_n$ are the zeros of the polynomial $P_n(u)$. A detailed study of these zeros permits one to reduce the large bracket to the form $x^n/q(x)$, where q(x) is a polynomial with real coefficients and such that $q(x) > 0$ at least for $x > -\min_{1 \leq j \leq n} |\alpha_j|^2$. This shows that $\phi(x) > 0$ for

$0 < x < \infty$; hence, $G(t) \geq 0$ on $0 < t < \infty$ and $\phi(x)$ is indeed completely monotonic, as we wanted to show.

PROBLEM 3. To solve the equation with partial derivatives

(1) $\Delta V = \frac{1}{c^2}\frac{\partial^2 V}{\partial t^2}$ where Δ is the Laplacian,

$$\Delta = \frac{\partial^2}{\partial r^2} + \frac{2}{r}\frac{\partial}{\partial r} + \frac{1}{r^2}\left(\frac{\partial^2}{\partial\theta^2} + \cot\theta\frac{\partial}{\partial\theta}\right) + \frac{1}{r^2\sin^2\theta}\frac{\partial^2}{\partial\phi^2},$$

with the boundary conditions (i) and (ii):

(i) $V = V(r,\theta,\phi,t)$ is symmetric with respect to a "polar axis" through the origin, so that, in fact, $V = V(r,\theta,t)$ only;

(ii) V is monochromatic, i.e., all "waves" have the same frequency ω; and with the initial condition

(iii) the values of V are prescribed along the polar axis at $t = 0$, say $V(r,0,0) = f(r)$, a given function of r.

Here r, θ, and ϕ are the customary spherical coordinates, t stands for the time and c represents dimensionwise a velocity.

Conditions (i) and (ii) are imposed only in order to simplify the problem and can be omitted, but the added complexity can easily be handled by classical methods and has nothing to do with the problem on hand.

Following the lead of Krall and Frink [68] we look, in particular, at solutions of (1) of the form (obtained by separation of variables)

$$u = r^{-1}y(1/ikr)L(\cos\theta)e^{ik(ct-r)}.$$

Here y and L are, so far, undetermined functions and we shall determine them precisely by the condition that u be a solution of (1), while k is a parameter related to the frequency ω by $k = \omega/c$. On account of (ii), k is a well defined constant. The (artificially looking) device of introducing complex elements into this physical problem is useful for obtaining propagating, rather then stationary waves. The real components v, w of $u = v + iw$ will be real solutions of (1) and represent waves traveling with the velocity c. For $c = 0$ and with $x = 1/ikr$, one obtains, of course, directly a real, stationary solution of (1). We now substitute u into (1), by taking into account that $\partial u/\partial\phi = 0$, and obtain, with $x = 1/ikr$ and $z = \cos\theta$, that

$$L(z)(x^2y''(x) + (2+2x)y'(x)) + y(x)((1-z^2)L''(z) - 2z\,L'(z)) = 0,$$

or, equivalently,

$$\frac{x^2y''(x)+(2+2x)y'(x)}{y(x)} = -\frac{(1-z^2)L''(z)-2zL'(z)}{L(z)}.$$

These two functions, each of which depends on a different independent variable, can be identically equal only if their common value reduces to a constant, say C. It follows that L(z) satisfies an equation of the form

$$(1-z^2)L''(z)-2z\,L'(z) + CL(z) = 0.$$

We immediately recognize here the classical equation of Legendre. If, but only if $C = n(n+1)$ with n an integer (it is clearly sufficient to consider only $n \geq 0$, because $-n(-n+1) = n(n-1)$) does this differential equation admit a polynomial solution, namely the Legendre polynomial of exact degree n; we shall denote it by $L_n(z)$. Incidentally, if we would not require symmetry with respect to the polar axis, then we would obtain here the associate Legendre polynomials $P_n^{(q)}(z)$ instead of the simpler Legendre polynomials, and this is the main reason for the present, more restrictive formulation of the problem.

So far, everything has been fairly routine; now, however, it turns out rather surprizingly that, with $C = n(n+1)$ the equation

$$x^2\,y''(x) + (2+2x)\,y'(x) - n(n+1)\,y(x) = 0, \tag{2}$$

satisfied by y(x), also admits a polynomial solution for n an integer, namely a polynomial of exact degree n, uniquely determined up to an arbitrary multiplicative constant. We shall denote it by $y_n(x)$ and may normalize it, e.g., by setting $y_n(0) = 1$.

We have, herewith, obtained a sequence of solutions to (1), of the form

$$u_n = u_n(r,\theta,t) = r^{-1}L_n(\cos\theta)y_n(1/irk)e^{ik(ct-r)}.$$

With each solution u_n and for each constant a_n, also a_nu_n is a solution of (1), and so is the sum $V = \sum_{n=0}^{\infty} a_nu_n$, if it converges. In particular, along the polar axis, with $z = \cos\theta = 1$, $L_n(1)$ is equal to 1, and we obtain at $t = 0$, with previous substitution $x = 1/ikr$,

$$V = V(r,0,0) = \sum_{n=0}^{\infty} a_n r^{-1}e^{-ikr}y_n(1/ikr) = ik\sum_{n=0}^{\infty} a_n xe^{-1/x}y_n(x).$$

In order to satisfy also the initial condition, we define F(x) by $f(r) = f(1/ikx) = ikx\,F(x)$, so that condition (iii) becomes

$$\sum_{n=0}^{\infty} a_ne^{-1/x}y_n(x) = F(x).$$

From (2) it follows that, by taking as closed path of integration the unit circle,

$$\oint y_n(z)y_m(z)e^{-2/z}dz = \delta_{mn}\frac{2(-1)^{n+1}}{2n+1},$$

where the Kronecker delta $\delta_{mn} = 1$ for $m = n$, $\delta_{mn} = 0$ otherwise. It follows that

$$a_n = (-1)^{n+1}(n+1/2)\oint F(z)y_n(z)e^{-1/z}dz.$$

With these values for a_n, $V(r,\theta,t) = \sum_{n=0}^{\infty} a_n r^{-1}L_n(\cos\theta)y_n(1/ikr)e^{ik(ct-r)}$ is a formal solution of (1), in general complex valued, that satisfies formally all boundary and initial conditions of the problem. It is an actual solution, if the infinite series converges. Precise conditions (that depend on the nature - especially the singularities - of F(z)) are known (see [13]) for this convergence and will be discussed in Chapter 9. Here we add only that in the more general situation, when we discard the restrictions (i) and (ii), the corresponding solution is of the form

$$V(r,\theta,\phi,t) = \sum_k \sum_{n=0}^{\infty} \sum_{m=0}^{n} a_{n,k}P_n^m(\cos\theta)\sin(m\phi+\phi_0)e^{ik(ct-r)}y_n(1/ikr)/r,$$

with the outer sum extended over all values of $k = \omega/c$, corresponding to all frequencies ω that occur.

What do these problems have in common? All three depend on the study of certain sequences of polynomials, $A_n(x)$ $(= -B_n(-x))$ in Problem 1, $P_n(x)$ in Problem 2, $y_n(x)$ in Problem 3. In fact, the three problems have more in common than just that, because actually, all three sequences of polynomials are essentially the same sequence. There are still many other problems, in which this particular sequence of polynomials known to-day as Bessel Polynomials plays a fundamental role. It also turns out that these polynomials exhibit certain symmetries that are esthetically appealing and have therefore been studied for their own sake. To-day there exists a fairly extensive literature devoted to this specific subject. Nevertheless, recently, when the present author needed some information concerning these polynomials, it turned out that it required an inordinate amount of time to search through a large number of papers and several books, in order to locate many a particular fact needed. It is the purpose of the present monograph, to give a coherent account of this interesting sequence of polynomials. It may be overly optimistic to hope that everything known about them will be found here, but at least the more important theorems will be stated and proved. Originally an attempt was made to obtain all important properties in a unified way, but this attempt has not always been successful; in fact it could hardly have been expected to be. After all, it is not surprizing that the structure of the Galois group of $P_n(x)$ requires for its determination other methods than, say, the study of the domain of convergence of an expansion in a series of these same polynomials.

The author has made himself a modest contribution to the subject matter, but the aim of this work is primarily expository: to systematize and to make easily accessible the work of all mathematicians active in this field. But mainly, unless this book succeeds in relieving the future student of this subject of the need for an exasperating, time consuming search for known items, deeply hidden in the literature, it will have failed in its purpose.

PART I

CHAPTER 1

HISTORIC SKETCH

It may not be easy to determine the first occurrence in the mathematical literature of the sequence of polynomials that we are interested in. They seem to have appeared sporadically for a rather long time. In 1873 Hermite [29] proved the transcendency of e. The polynomials used by Hermite are closely related to the object of our study. See [83] for the connection. The polynomials denoted by Olds with T_n and Z_n are related to the Bessel Polynomials $y_n(z)$ by $y_n(z) = T_n(z) + Z_n(z)$; see [83] and Chapter 8 for more details.

These polynomials appear also in 1929, in a paper by Bochner [14] and in one by Romanowsky [92]. Shortly afterwards (1931), but quite independently, they occur in a long paper [18] by J.L. Burchnall and T.W. Chaundy, who obtain them as solutions of certain differential equations. W. Hahn [58] runs across them in 1935 and so did H.L. Krall [67] in 1941.

The first systematic study of these polynomials is due to H.L. Krall and O. Frink, who in 1949 consider them in a fundamental paper [68] published in the Transactions of the AMS. They gave these polynomials the name of BESSEL POLYNOMIALS, under which they have been known ever since. This same designation has often been used by various authors, even when they actually studied related polynomials, or different normalizations, etc. So, e.g., we find among the sets of polynomials labelled as Bessel Polynomials, besides Krall and Frink's $y_n(x)$, also $x^n y_n(1/x)$, $(-1)^{n-1} x^n y_n(2/x)$, $(x/2)^n y_n(2/x)$, and others.

In the present monograph, we shall adopt in general the original normalization of Krall and Frink and we shall abbreviate the designation Bessel Polynomials, by BP (regardless of their use in the singular, or plural).

Krall and Frink had been led to the consideration of the BP by a study of the wave equation (essentially our Problem 3). They indicate the differential equation, recurrence relations, a pseudo generating function and an orthogonality property; they also generalize the set $y_n(x)$, by introducing two parameters (only one really significant).

Shortly afterwards, and stimulated by Krall and Frink's work, two other papers appeared almost simultaneously. J.L. Burchnall [17] pointed out that the BP $y_n(x)$ is related to the polynomials $\theta_n(x)$ studied by Burchnall and Chaundy in [18] by $\theta_n(x) = x^n y_n(1/x)$. By using the machinery developed there, Burchnall obtains several of the results of Krall and Frink and, in addition some of the beautiful symmetry properties of the zeros, as well as a generating function.

The present author [53] studied asymptotic properties of the BP, also properties of their zeros, the irreducibility of the BP over the rationals and the Galois group of the BP.

At about the same time, when Krall and Frink, Burchnall and the present author started the systematic study of BP, W.E. Thomson studied certain networks used in multistage amplifiers, and that led to a particularly desirable characteristic, called maximally - flat delay. The investigation of the complex impedance and of the transfer functions of these networks led to the consideration of certain polynomials, proportional to the transfer function and defined by initial values and a recurrence relation. These polynomials turn out to be exactly the BP in the normalization of Burchnall and Chaundy. It is quite unfortunate that Thomson's work [107], [108] remained essentially unknown to the mathematicians who worked on BP. Indeed, Thomson obtained many important properties of the zeros of BP, but he was not particularly interested in the theory. "Those interested in the theory will find an outline in the Appendix", he writes. It is indeed just an outline, set in particularly small print. In addition to many interesting details, some of which were rediscovered only recently (see Chapter 10),the paper [108] contains also a tabulation of all the zeros (real and complex) of the BP up to the ninth degree (inclusive). For other similar tabulations see [99], [100], [70], [102], [69], and [116].

A few years afterwards, a real flood of papers appeared, with improvements of the theorems concerned with the location of the zeros, with generating functions and with the relations of the BP to other special functions, especially to Bessel functions and the hypergeometric series; this is not too surprizing, since $y_n(x) = {}_2F_0(-n, 1+n; -; -x/2)$. It is impossible to mention at this place all contributions made during these years, and quoted in the following pages but a few names and papers come to mind: Carlitz [19]; Al-Salam and Carlitz [7], [8], Al-Salam [3], Rainville [89], Agarwal [2]; Toscano [109]; McCarthy [74], [75]; Nasif [81], Dickinson [47], Brafman [15], Ragab [88], Wimp [112] and others.

Some of this material was included in abbreviated form in the books by R.P. Boas and R.C. Buck [13] and by E.D. Rainville [90].

In 1962 Dočev [48] obtained what until recently was the best upper bound for the absolute value of the zeros of BP. See Chapter 10 for more recent results by Olver [84], [46] and by Saff and Varga [98].

While the interest in BP never completely disappeared, there were a few years, during which the efforts of mathematicians were apparently directed into other channels. But recently, there seems to have arisen a renewed interest in BP. In 1969 the author [54] could settle (computer assisted - although in the proof itself no computer work is invoked) a remaining problem concerning the Galois groups of the BP of degrees 9, 11, and 12.

Parodi [85] represents the BP as determinants and obtains new bounds for the region in which one may find zeros of BP. If one combines these with some previously known ones, one gets rather strong results.

Barnes [12] studies again the zeros of BP, as well as their connection with continued fractions and with the exponential function.

Wragg and Underhill [113] relate the BP to the denominators of the Padé approximants to the exponential function; they also represent, following Parodi, the BP as determinants and, by use of Gershgorin's and the Bendixson-Hirsch theorems, obtain rather simple proofs for good bounds on the zeros of the BP.

Kelker [65] and Ismail and Kelker [62] reduce an important problem of probability (the infinite divisibility of certain distributions - essentially our Problem 2) to the complete monotonicity of the function $\phi(x) = y_{n-1}(x^{-1/2})/x^{1/2}y_n(x^{-1/2})$. The proof that this property, in fact, holds, has been obtained by the present writer [55], [56].

In the meantime, Thomson's contribution, ignored by mathematicians, became rapidly common knowledge among electrical engineers. They discovered soon the identity of Thomson's polynomials with the BP and used systematically the properties of these polynomials and of their zeros, as presented in [68], [17], and [53], in order to perfect amplifiers, as well as filters. In 1954, L. Storch [106a,b] giving full credit to Thomson, and quoting [108], [68], [17] and [53] treats the topic with all the details needed for the understanding of the theory and also for the effective computation of the numerical values of the elements of the network.

Soon the use of BP in the synthesis of certain networks was treated in textbooks like those of E. Guillemin [57] and D. Hazony [60]. More recently R.R. Shepard [101] indicated an almost mechanical method for the design of networks with certain preassigned characteristics, by use (among others) of BP. For related work see also [64], [115], and [116].

At present, the BP are accepted along with the classical orthogonal polynomials among the "special functions" and are often mentioned in connection with either the Bessel functions, or some other previously studied functions, to which they are reducible (see Chapter 5). They are quoted, besides in [90], also in such well-known collections as [71], [72], [20], [50] or [114](but not in [1]), and there does not appear to exist in the mathematical literature any systematic discussion of their properties. This fact, the author hopes, will be accepted as a sufficient justification for the present monograph.

CHAPTER 2

BESSEL POLYNOMIALS AND BESSEL FUNCTIONS:

DIFFERENTIAL EQUATIONS AND THEIR SOLUTIONS.

1. The so called "modified" Bessel functions $I_\nu(z)$ and $K_\nu(z)$ - sometimes improperly designated as being of imaginary argument -, $K_\nu(z)$ also known as the MacDonald function, satisfy the well-known (see e.g. [61] or [1]) differential equation

(1) $$Lw \equiv z^2w'' + zw' - (z^2 + \nu^2)w = 0.$$

For fixed ν with $\text{Re}\ \nu > 0$ the following asymptotic relations hold (see [1]): For $z \to 0$,

(2a) $$K_\nu(z) \cong \frac{1}{2}\Gamma(\nu)(z/2)^{-\nu},\ K'_\nu(z) \cong -\Gamma(\nu+1)2^{\nu-1}z^{\nu-1}$$
$$I_\nu(z) \cong \{\Gamma(\nu+1)\}^{-1}(z/2)^\nu,\ I'_\nu(z) \cong (2\Gamma(\nu))^{-1}(z/2)^{\nu-1};$$

for $z \to \infty$ and $|\arg z| < \frac{\pi}{2}$,

(2b) $$K_\nu(z) \cong (\pi/2z)^{1/2}e^{-z},\ I_\nu(z) \cong (2\pi z)^{-1/2}e^z.$$

These relations suggest that functions like $\phi(z)$, or $\theta(z)$, defined by $w = z^{-\nu}\phi$, or by $w = z^{-\nu}e^{-z}\theta$, with $\theta = e^z\phi$, may exhibit a simpler behavior than the Bessel functions themselves.

2. Let us consider first $\theta = \theta(z)$. By logarithmic differentiation

$$w' = (-\frac{\nu}{z} - 1 + \frac{\theta'}{\theta})w$$

$$w'' = (\frac{\nu}{z^2} + \frac{\theta''}{\theta} - (\frac{\theta'}{\theta})^2)w + (-\frac{\nu}{z} - 1 + \frac{\theta'}{\theta})w'$$

$$= [\frac{\nu}{z^2} + \frac{\theta''}{\theta} - (\frac{\theta'}{\theta})^2 + (-\frac{\nu}{z} - 1 + \frac{\theta'}{\theta})^2]w$$

and, substituting these in (1) we obtain

$$Lw = (z\theta'' - (2z+2\nu-1)\theta' + (2\nu-1)\theta)zw\theta^{-1} = 0.$$

As $zw(z) \neq 0$, it follows that $\theta(z)$ satisfies the differential equation

(3) $$z\theta'' - (2z+2\nu-1)\theta' + (2\nu-1)\theta = 0.$$

By the general theory of linear differential equations (e.g., [18]) the origin is a regular singular point and there exist (in general) two independent particular solutions of the form $\theta = z^\alpha \sum_{m=0}^{\infty} c_m z^m$. Here α is any one of the two (in general distinct) solutions of the indicial equation. Either by Frobenius' method, or by general considerations one finds that the indicial equation is

(4) $$\alpha(\alpha - 2\nu) = 0.$$

The solutions of (4) are indeed distinct, except for $\nu = 0$. In the latter case the two solutions are, of course, proportional to $e^{-z}I_0(z)$ and $e^{-z}K_0(z)$, respectively and are of no particular further interest here.

For $\nu \neq 0$, set $\theta = \theta(z,\nu) = \sum_{m=0}^{\infty} c_m z^m$ and $\tilde{\theta} = \tilde{\theta}(z,\nu) = z^{2\nu} \sum_{m=0}^{\infty} c'_m z^m$. By differentiation and substitution into (3), we obtain

$$c_m = \frac{2\nu+1-2m}{(2\nu-m)m} c_{m-1} = \dots = \frac{(2\nu-1)(2\nu-3)\dots(2\nu-2m+1)}{(2\nu-1)(2\nu-2)\dots(2\nu-m)m!} c_0 ,$$

so that

$$\theta(z,\nu) = c_0(1+z+ \frac{(2\nu-1)(2\nu-3)}{(2\nu-1)(2\nu-2)} \frac{z^2}{2!} + \dots + \frac{(2\nu-1)(2\nu-3)\dots(2\nu-2m+1)}{(2\nu-1)(2\nu-2)\dots(2\nu-m)} \frac{z^m}{m!} + \dots).$$

Similarly,

$$c'_m = \frac{2\nu+2m-1}{(2\nu+m)m} c'_{m-1} = \dots = \frac{(2\nu+1)(2\nu+3)\dots(2\nu+2m-1)}{(2\nu+1)(2\nu+2)\dots(2\nu+m)m!} c'_0$$

and

$$\tilde{\theta}(z,\nu) = c'_0 z^{2\nu}(1+z+ \frac{2\nu+3}{2\nu+2} \frac{z^2}{2!} + \dots + \frac{(2\nu+1)(2\nu+3)\dots(2\nu+2m-1)}{(2\nu+1)(2\nu+2)\dots(2\nu+m)} \frac{z^m}{m!} + \dots) .$$

The functions $\theta(z,\nu)$ and $z^\nu e^z K_\nu(z)$, both are solutions of (3) and so is any linear combination with constant coefficients of these functions. Also, relations (2a) show that if we set $u_\nu(z) = c_0 z^\nu e^z K_\nu(z) - 2^{\nu-1}\Gamma(\nu)\theta(z,\nu)$, then $u_\nu(z)$ is a solution of (3) with $u_\nu(0) = u'_\nu(0) = 0$. Consequently, $u_\nu(z)$ vanishes identically and

(5) $$\theta(z,\nu) = \mu_\nu z^\nu e^z K_\nu(z) \text{ with } \mu_\nu = c_0 2^{1-\nu}(\Gamma(\nu))^{-1}.$$

In a similar way one verifies that $\tilde{\theta}(z,\nu) = \tilde{\mu}_\nu z^\nu e^z I_\nu(z)$, with $\tilde{\mu}_\nu = c'_0 2^\nu \Gamma(\nu+1)$.

It is obvious that $\theta(z,\nu)$ reduces to a polynomial of exact degree n if $2\nu = 2n+1$ and that $z^{-2\nu}\,\tilde{\theta}(z,\nu)$ reduces to a polynomial of exact degree n if $2\nu = -2n-1$. In the first case, for $\nu = n+1/2$, the two independent solutions of (3) are

$$\theta(z,n+\tfrac{1}{2}) = c_0(1+z+ \frac{2n-2}{2n-1} \frac{z^2}{2!} + \dots + \frac{(2n-2)(2n-4)\dots(2n-2m+2)}{(2n-1)(2n-2)\dots(2n-m+1)} \frac{z^m}{m!} + \dots + \frac{(2n-2)(2n-4)\dots4\cdot2}{(2n-1)(2n-2)\dots(n+2)(n+1)} \frac{z^n}{n!})$$

and

$$\tilde{\theta}(z,n+\tfrac{1}{2}) = c_0' z^{2n+1}\left(1+z+\frac{2n+4}{2n+3}\frac{z^2}{2!}+\ldots+\frac{(2n+4)(2n+6)\ldots(2n+2m)}{(2n+3)(2n+4)\ldots(2n+m+1)}\frac{z^m}{m!}+\ldots\right).$$

In the second case, with $\nu = -n - \frac{1}{2}$,

$$\theta(z,-n-1/2) = c_0\left(1+z+\ldots+\frac{(2n+4)(2n+6)\ldots(2n+2m)}{(2n+3)(2n+4)\ldots(2n+m+1)}\frac{z^m}{m!}+\ldots\right)$$

and

$$\tilde{\theta}(z,-n-1/2) = c_0' z^{-2n-1}\left(1+z+\ldots+\frac{(2n-2)(2n-4)\ldots(2n-2m+2)}{(2n-1)(2n-2)\ldots(2n-m+1)}\frac{z^m}{m!}+\ldots\right.$$

$$\left.+\frac{(2n-2)\ldots 2}{(2n-1)\ldots(n+1)}\frac{z^n}{n!}\right).$$

If we take $c_0 = c_0'$, then

(6) $\qquad \theta(z,n+1/2) = z^{2n+1}\tilde{\theta}(z,-n-1/2)$ and $\tilde{\theta}(z,n+1/2) = z^{2n+1}\theta(z,-n-1/2)$.

It follows that, in general, it will be sufficient to consider only non-negative values for the half integral second parameter. Sometimes, however, e.g., for reasons of symmetry, it may be convenient to be allowed to use also negative values for the integer n and we shall soon find a convenient way to do it.

In $\theta(z,n+1/2)$ the coefficient of z^n equals $c_0\cdot 2^n n!/(2n)!$. In following Burchnall [17] and Burchnall and Chaundy [18], we select $c_0 = (2n)!/2^n n!$, so that the leading coefficient becomes one. With this normalization we shall denote $\theta(z,n+1/2)$ simply by $\theta_n(z)$. One now easily verifies (see [17]) that

(7) $$\theta_n(z) = \sum_{m=0}^{n} a_{n-m} z^m = \sum_{m=0}^{n} a_m z^{n-m},$$

where all coefficients are integers (see Theorem 3.1) given by

(8) $$a_{n-m} = \frac{2^{m-n}(2n-m)!}{(n-m)!m!}, \quad a_m = \frac{(n+m)!}{2^m(n-m)!m!}.$$

So, e.g., we find that

$$\theta_0(z) = 1,\ \theta_1(z) = z+1,\ \theta_2(z) = z^2+3z+3,\ \text{etc.}$$

If we have to deal simultaneously with several polynomials, it is useful to identify the polynomials to which a coefficient belongs by a superscript. So, e.g., $a_3^{(5)}$ is the coefficient of $z^{5-3} = z^2$ in $\theta_5(z)$; by (8) its value is $\frac{(5+3)!}{2^3(5-3)!3!} = \frac{8!}{8\cdot 2!3!} = \frac{7!}{12} = 420$.

The coefficients that occur in (7) and (8) would then be denoted by $a_{n-m}^{(n)}$ and $a_m^{(n)}$, respectively.

While the polynomials $\theta_n(z)$ have been defined so far only for $n \geq 0$, we observe that, if we replace formally n by -n in (8), we obtain

$$a_m^{(-n)} = \frac{(-n-m+1)(-n-m+2)\dots(m-n)}{2^m m!} = (-1)^{2m} \frac{(n+m-1)(n+m-2)\dots(n-m)}{2^m m!} = a_m^{(n-1)}.$$

Hence, if we extend the definition (7) of $\theta_n(z)$ formally to negative subscripts, we obtain

$$(9) \qquad \theta_{-n}(z) = \sum_{m=0}^{n} a_m^{(-n)} z^{-n-m} = z^{-2n+1} \sum_{m=0}^{n} a_m^{(n-1)} z^{(n-1)-m}$$

$$= z^{-2n+1} \sum_{m=0}^{n-1} a_m^{(n-1)} z^{(n-1)-m} = z^{-2n+1} \theta_{n-1}(z);$$

indeed, on account of the factor n-m in the numerator of $a_m^{(n)}$, the coefficient $a_n^{(n-1)}$ vanishes.

The identity (9) may be taken as definition of $\theta_{-n}(z)$. If we replace in (8) n by n+1, we obtain $\theta_n(z) = z^{2n+1}\theta_{-(n+1)}(z)$. Comparison with (6) shows that $\theta_{-(n+1)}(z)$ is not a new function, but is precisely the function $\tilde{\theta}(z,-n-1/2)$.

In some contexts it is more convenient to work with the reverse polynomial

$$(10) \qquad y_n(z) = z^n \theta_n(1/z) = \sum_{m=0}^{n} a_m z^m,$$

a normalization due to Krall and Frink (see [68]). If we want to have the first coefficient reduced to unity, we may factor out $a_n^{(n)} = ((2n)!)/2^n n! = c_0$ and obtain

$$y_n(z) = c_0 \sum_{m=0}^{n} b_m^{(n)} z^{n-m}, \quad b_m^{(n)} = \frac{2^m}{m!} \quad \frac{(2n-m)(2n-m-1)\dots(n-m+1)}{2n(2n-1)\dots(n+1)}.$$

By differentiation of $\theta_n(z) = z^n y_n(1/z)$ we obtain successively

$$\theta_n'(z) = nz^{n-1}y_n(1/z) - z^{n-2}y_n'(1/z),$$

and

$$\theta_n''(z) = n(n-1)z^{n-2}y_n(1/z) - nz^{n-3}y_n'(1/z) - (n-2)z^{n-3}y_n'(1/z) + z^{n-4}y_n''(1/z).$$

If we substitute these into (3) with $\nu = n+1/2$, we obtain after a few simplifications (see [68]) that $y_n(x)$ satisfies the differential equation

$$(11) \qquad z^2 y_n''(z) + 2(z+1)y_n'(z) - n(n+1)y_n(z) = 0.$$

From (9) and (10) it follows that

$$y_n(z) = z^n\theta_n(z^{-1}) = z^n \cdot z^{-2n-1}\theta_{-(n+1)}(z^{-1}) = z^{-n-1}\theta_{-(n+1)}(z^{-1}) = y_{-n-1}(z),$$

so that the polynomials $y_n(z)$ are defined for negative subscripts by the particularly simple relation (see [68] and [47])

$$y_{-n}(z) = y_{n-1}(z). \tag{12}$$

We shall speak occasionally of both, $\theta_n(z)$ and $y_n(z)$ as Bessel Polynomials (BP) and it will be clear from the context, which one is meant. When confusion could arise, then $y_n(z)$ will be called the n-th BP, while $\theta_n(z)$ will be referred to (following Boas and Buck [13]) as the reverse BP.

We summarize (and slightly complete) the results obtained in this section in the following theorem:

THEOREM 1: *For integral* $n > 0$, *the differential equation*

$$zw'' - 2(z+n)w' + 2nw = 0$$

has polynomial solutions $w = \theta_n(z)$. *These are polynomials of exact degree* n *and specifically* $\theta_n(z) = \sum_{m=0}^{n} a_m^{(n)} z^{n-m}$, *with* $a_m^{(n)} = \dfrac{(n+m)!}{2^m(n-m)!m!}$.

If $n = 0$, *the general solution of the differential equation is* $c_1e^{2z} + c_2$ *and also in this case among the solutions one finds the polynomial of degree zero* $\theta_0(z) = 1$, *the same as one obtains formally by setting* $n = 0$ *in above formulae.*

The equation with n *replaced by the negative integer* -n, $n > 0$ *has a rational solution* $\theta_{-n}(z)$ *that satisfies* $\theta_{-n}(z) = z^{-2n+1}\theta_{n-1}(z)$, *or, equivalently,*

$$\theta_n(z) = z^{2n+1}\theta_{-(n+1)}(z).$$

The equation

$$z^2w'' + 2(z+1)w' - n(n+1)w = 0$$

has polynomial solutions $w = y_n(z)$ *of exact degree* n. *Specifically, with the same* $a_m^{(n)}$ *as above,* $y_n(z) = \sum_{m=0}^{n} a_m^{(n)} z^m$ *is a solution, so that* $y_n(z) = z^n\theta_n(1/z)$ *and* $y_n(z)$ *and* $\theta_n(z)$ *are polynomials reverse to each other.*

For $n = 0$ *the general solution of the equation is* $c_1e^{2/z} + c_2$ *and among these solutions one has, in particular, the polynomial of degree zero* $y_0(z) = 1$.

If n *is replaced by* -n-1, *the product* n(n+1) *remains unchanged, so that*

$$y_{-n-1}(z) = y_n(z).$$

3. We turn now to the function $\phi(z) = e^{-z}\theta(z)$. From $\theta(z) = e^{z}\phi(z)$, by differentiation, $\theta' = e^{z}\phi' + e^{z}\phi$, $\theta'' = e^{z}\phi'' + 2e^{z}\phi' + e^{z}\phi$ and if we substitute these in (3), it follows that ϕ satisfies the particularly simple differential equation

(13) $$z^2\phi'' - 2nz\phi' = z^2\phi.$$

Denote the linear differential operator $z^2 \frac{d^2}{dz^2} - 2nz \frac{d}{dz}$ by L; then the equation

$$L\phi = z^2\phi$$

stays invariant under the transformation $z \to -z$. Hence, with $\phi(z) = e^{-z}\theta(z)$, also $\phi(-z) = e^{z}\theta(-z)$ will be a solution of (13).

The differential operator L, of second order, can be factored with the help of the first order operator $\delta = z \frac{d}{dz}$. For future use, and also because they are of independent interest, some of the (well known - see [17]) properties of this operator will be developed here.

Clearly $\delta z^n = nz^n$

and, more generally,

(14) $$\delta^k z^n = n^k z^n$$

holds, and also

(15) $$(\delta-a)z^n = (n-a)z^n.$$

More generally, if f(z) is differentiable, $\delta f = zf'$ and, if f(z) is n times differentiable, then

$$\delta^n f = \sum_{m=1}^{n} S_n^{(m)} z^m f^{(m)},$$

where $S_n^{(m)}$ are the Stirling numbers of the second kind. The general case follows by induction on n, starting from n = 1 (which holds by the definition of δ), with the help of the recurrence relation (see [1], p. 825) $S_{n+1}^{(m)} = mS_n^{(m)} + S_n^{(m-1)}$.

Next, by using (15) one verifies that the operators δ-a and δ-b commute. Consequently, if P(n) is a polynomial with zeros $u_1, u_2, \ldots, u_m$, one has

(16) $$P(\delta)z^k = (k-u_1)(k-u_2)\ldots(k-u_m)z^k = P(k)\cdot z^k.$$

We observe that for every differentiable function f(z),

$$(\delta-n)zf(z) = z\frac{d}{dz}(zf(z))-nzf(z) = z(zf'-(n-1)f) = z(\delta-(n-1))f(z).$$

It follows that a factor z may be "moved across" the operator $\delta-n$, from right to left, provided that n is replaced by n-1.

If we iterate the procedure we obtain the result that

$$(\delta-n_1)(\delta-n_2)\dots(\delta-n_r)zf(z) = z(\delta-n_1+1)\dots(\delta-n_r+1)f(z). \tag{17}$$

In particular, for $f(z) = e^{-bz}$, we have

$$(\delta-n_1)(\delta-n_2)\dots(\delta-n_r)ze^{-bz} = z(\delta-n_1+1)(\delta-n_2+1)\dots(\delta-n_r+1)e^{-bz}. \tag{18}$$

It is possible to compute explicitly the right hand side of (18).

This has a particularly simple expression, if the constants $n_1, n_2, \dots, n_r$ are consecutive integers.

Indeed,

$$(\delta-n)e^{-bz} = -(bz+n)e^{-bz} = z^{n+1}\frac{d}{dz}(z^{-n}e^{-bz}),$$

and

$$\delta(\delta-n)e^{-bz} = (n+1)z^{n+1}\frac{d}{dz}(z^{-n}e^{-bz})+z^{n+2}\frac{d^2}{dz^2}(z^{-n}e^{-bz}),$$

so that $(\delta-n-1)(\delta-n)e^{-bz} = z^{n+2}\dfrac{d^2}{dz^2}(z^{-n}e^{-bz})$.

An induction on m will now complete the proof of

$$(\delta-n-m+1)(\delta-n-m+2)\dots(\delta-n)e^{-bz} = z^{n+m}\frac{d^m}{dz^m}(z^{-n}e^{-bz}). \tag{19}$$

We observe in particular that for a twice differentiable function f(z),

$$\delta(\delta-a)f = z^2f'' + (1-a)zf',$$

and if, in particular, $a = 2n+1$, then $\delta(\delta-2n-1)f = Lf$. This yields the announced factorization of L. From (13) it now follows, that $\phi(z)$ is a solution of

$$\delta(\delta-2n-1)\phi = z^2\phi. \tag{20}$$

THEOREM 2. (Burchnall [17], Chaundy and Burchnall [18]). *Define the differential operator* $Q_n(\delta) = (\delta-1)(\delta-3)\dots(\delta-(2n-1))$; *then the function* $\phi_n(z) = Q_n(\delta)e^{-z}$ *is a solution of* (20).

Proof. $\delta(\delta-2n-1)\phi_n = \delta(\delta-2n-1)(\delta-1)(\delta-3)\dots(\delta-2n+1)e^{-z}$

$$= (\delta-1)(\delta-3)\dots(\delta-2n-1)\delta e^{-z} = -(\delta-1)(\delta-3)\dots(\delta-2n-1)ze^{-z}$$

$$= -z\delta(\delta-2)\dots(\delta-2n)e^{-z} = -z(\delta-2)(\delta-4)\dots(\delta-2n)\delta e^{-z}$$

$$= z(\delta-2)(\delta-4)\dots(\delta-2n)ze^{-z} = z^2(\delta-1)(\delta-3)\dots(\delta-2n+1)e^{-z} = z^2\phi_n.$$

Here the first and last equality are justified by the definition of $\phi_n(z)$, the second and fifth one by the commutativity of the operators $\delta-a$, the third and sixth one by the definition of δ, and the fourth and seventh by (18). Theorem 2 may also be proved directly, as follows:

By using (16),

$$(21) \qquad \phi_n(z) = Q_n(\delta)e^{-z} = Q_n(\delta)\sum_{k=0}^{\infty}(-1)^k\frac{z^k}{k!} = \sum_{k=0}^{\infty}\frac{(-1)^k}{k!}Q_n(\delta)z^k$$

$$= \sum_{k=0}^{\infty}\frac{(-1)^k}{k!}Q_n(k)z^k$$

and

$$(22) \qquad z^2\phi_n(z) = \sum_{k=0}^{\infty}\frac{(-1)^k}{k!}Q_n(k)z^{k+2} = \sum_{k=2}^{\infty}\frac{(-1)^k}{(k-2)!}Q_n(k-2)z^k.$$

Next,

$$\delta\phi_n(z) = \sum_{k=0}^{\infty}\frac{(-1)^k}{k!}Q_n(k)\delta z^k = \sum_{k=0}^{\infty}(-1)^k\frac{k}{k!}Q_n(k)z^k = \sum_{k=1}^{\infty}\frac{(-1)^k}{(k-1)!}Q_n(k)z^k$$

$$= \sum_{k=2}^{\infty}\frac{(-1)^k}{(k-2)!}(k-3)(k-5)\dots(k-2n+1)z^k,$$

and similarly,

$$\delta^2\phi_n(z) = \sum_{k=2}^{\infty}\frac{(-1)^k}{(k-2)!}k(k-3)\dots(k-2n+1)z^k.$$

Hence,

$$(23) \qquad L\phi_n = (\delta^2-(2n+1)\delta)\phi_n = \sum_{k=2}^{\infty}\frac{(-1)^k}{(k-2)!}(k-3)(k-5)\dots(k-(2n+1))z^k$$

$$= \sum_{k=2}^{\infty}\frac{(-1)^k}{(k-2)!}Q_n(k-2)z^k = z^2\phi_n,$$

by (22).

For later use we state the following.

THEOREM 3. *In the series expansion of* $\phi_n(z)$, *the coefficients of* z^k *vanish for* $k = 1,3,5,\dots,(2n-1)$.

Proof. $Q_n(k) = 0$ for $k = 1,3,\ldots,(2n-1)$, so that Theorem 3 follows from (21).

As already observed, (13) and, hence, (20), are satisfied not only by $\phi_n(z)$, but also by $\phi_n(-z)$. This is evident also from (22) and (23), in both of which one only has to suppress the factor $(-1)^k$ under the summation sign.

From $\phi_n(z) = Q_n(\delta)e^{-z}$ and the relation between $\theta_n(z)$ and $\phi_n(z)$ it follows that

$$\theta_n(z) = e^z Q_n(\delta)e^{-z},$$

an identity found already in [17].

For future use it is of interest to record here also the δ-forms of the differential equations for $\theta_n(z)$ and $y_n(z)$:

(24) $$[\delta(\delta-2n-1)-2z(\delta-n)]\theta_n(z) = 0 ,$$

(25) $$[2\delta + z(\delta-n)(\delta+n+1)]y_n(z) = 0.$$

If one replaces δ by $z\,\frac{d}{dz}$, one immediately verifies that these are, indeed, the same as (3) and (11), respectively.

4. Equation (11) has been generalized in several ways. Krall and Frink [68] introduce two new parameters and write the equation as

(26) $$z^2y''+(az+b)y'-n(n+a-1)y = 0.$$

For $a = b = 2$, (26) reduces, of course, to (11). Following [68], we shall denote the polynomial solution of (26) (if such exists) by $y_n(z;a,b)$, the generalized BP of degree n. The reverse BP of degree n, $z^n y_n(z^{-1};a,b)$ will be denoted by $\theta_n(z;a,b)$.

If we set $Y(z) = y(2z/b)$ and $Z = 2z/b$, then $Y'(z) = (2/b)y'(2z/b)$, $Y''(z) = (2/b)^2y''(2z/b)$ and we verify that

$$z^2Y''+(az+b)Y'-n(n+a-1)Y = (2z/b)^2y''(2z/b)+(a\,\frac{2z}{b} + 2)y'(2z/b)-n(n+a-1)y(2z/b)$$

$$= Z^2y''(Z)+(aZ+2)y'(Z)-n(n+a-1)y(Z).$$

This shows that if $y(Z)$ is a solution of (26) with $b = 2$, then $y(2z/b)$ is a solution of the general equation (26). In other words, b is only a scale factor for the independent variable and not an essential parameter. If we want to keep $b = 2$, but let $y_n(z)$ or $\theta_n(z)$ depend also on the parameter a, we write $y_n(z;a)$ and $\theta_n(z,a)$, respectively.

By Frobenius' method or by direct verification through substitution, it may be shown that

(27) $$y_n(z;a) = \sum_{k=0}^{n} d_k^{(n)} z^k, \text{ where } d_k^{(n)} = \frac{n!(n+k+a-2)^{(k)}}{k!(n-k)!2^k}.$$

In (27) and hereafter we use the notation $u^{(n)}$ to mean $u(u-1)\dots(u-n+1)$. Similarly, u_n will stand for $u(u+1)\dots(u+n-1)$. No confusion should arise between these standard notations and sub, or superscripts, like in $d_k^{(n)}$.

From these the coefficients $f_k^{(n)}$ of $y_n(z;a,b) = \sum_{k=0}^{n} f_k^{(n)} z^k$ are obtained, by replacing z by 2z/b; hence,

(28) $$f_k^{(n)} = \frac{n!(n+k+a-2)^{(k)}}{k!(n-k)!b^k}.$$

One verifies that, for a = b = 2, $f_k^{(n)} = \frac{n!(n+k)^{(k)}}{k!(n-k)!2^k} = \frac{(n+k)!}{2^k k!(n-k)!} = a_k^{(n)}$,

in agreement with (8).

Some authors like Obrechkoff [82] and Dočev [48] have adopted the normalizations $b = 1$, or $b = -1$ rather than $b = 2$. They also set $m = a-2$ and write m as a superscript. In their notation, therefore, $y_n(z;a) = (-1)^n P_n^{(a-2)}(-x/2)$ and, in particular, $y_n(z) = (-1)^n P_n^{(0)}(-x/2)$.

5. The corresponding generalization of $\theta_n(z)$ is obtained most conveniently by setting

$$y_n(z;a,b) = z^n \theta_n(z^{-1};a,b).$$

If we substitute this in (26), we obtain by routine computations the differential equation satisfied by $\theta_n = \theta_n(z;a,b)$, namely

(29) $$z\theta_n'' - (2n-2+a+bz)\theta_n' + bn\theta_n = 0.$$

For a = b = 2, (29) reduces to

$$z\theta_n'' - 2(z+n)\theta_n' + 2n\theta_n = 0,$$

which is, of course, (3) with $2\nu - 1 = 2n$.

The obvious generalization of $\phi_n(z)$ is clearly $\phi_n(z;a,b) = e^{-bz/2}\theta_n(z;a,b)$. It is convenient, however, to increase the flexibility of the presentation, by introducing a new parameter; hence, we set

(30) $$\phi_n(z;a,b,c) = e^{-cz}\theta_n(z;a,b).$$

From this we obtain as particular cases $\theta_n(z;a,b)$ itself, for $c = 0$, and $\phi_n(z;a,b)$ for $c = b/2$; in addition also the case $c = b$ will turn out to be of interest.

If we differentiate (30) twice and use (29) we obtain the differential equation satisfied by $w = \phi_n(z;a,b,c)$, namely

$$zw''-(2n+a-2+(b-2c)z)w'+(c(c-b)z+(b-2c)n+c(2-a))w = 0. \tag{31}$$

For $a = b = 2$, $c = 0$, (31) reduces again to (3) with $2\nu-1 = 2n$, while for $b = 2c = 2$, (31) becomes

$$zw''-(2n+a-2)w'+(2-a-z)w = 0, \tag{32}$$

satisfied by $\phi_n(z;a,2) = e^{-z}\theta_n(z;a,2)$.

Finally, for $b = c$, (31) simplifies to

$$zw''-(2n+a-2-bz)w'+b(2-a-n)w = 0, \tag{33}$$

with the solution $w = e^{-bz}\theta_n(z;a,b)$.

As in the particular case $a = b = 2c = 2$, it is convenient to "factor" (31) with the help of the first order differential operator $\delta = z\frac{d}{dz}$. From the definition of δ follows

$$\delta(\delta+1-a-2n)w = z\frac{d}{dz}\{zw'+(1-a-2n)w\} = z(zw''+(2-a-2n)w').$$

By (31) this equals

$$z^2(b-2c)w'+z\{c(b-c)z+(2c-b)n+(a-2)c\}w. \tag{34}$$

If $b \neq 2c$, (34) may be written as $z(b-2c)(\delta-n+\alpha z+\beta)w$, with $\alpha = c(b-c)/(b-2c)$, $\beta = c(a-2)/(b-2c)$, so that $w = \phi_n(z;a,b,c)$ is the solution of the differential equation

$$\delta(\delta+1-a-2n)w = -z(2c-b)(\delta-n+\alpha z+\beta)w. \tag{35}$$

For $b = 2c$, (34) equals $\{z^2(b/2)^2+(a-2)(b/2)z\}w$ and $w = \phi_n(z;a,b)$ satisfies

$$\delta(\delta+1-a-2n)w = \{z^2(b/2)^2+(a-2)(b/2)z\}w.$$

In particular, if $b = 2$, this reduces to

$$\delta(\delta+1-a-2n)w = (z^2+(a-2)z)w,$$

the δ-form of (32). If we set here also $a = 2$, we recover, of course, once more (20).

6. For b = c, (35) becomes (see (20) in [17])

$$(36)\qquad \delta(\delta+1-a-2n)w = -bz(\delta-n+2-a)w,$$

verified by $\phi_n(z;a,b,b) = e^{-bz}\theta_n(z;a,b)$.

In particular, for a = 2, (36) reduces to

$$(37)\qquad \delta(\delta-2n-1)w = -bz(\delta-n)w,$$

with the solution $w = e^{-bz}\theta_n(z;2,b)$.

Without restricting ourselves to a = 2, let us assume nevertheless (following Burchnall [17]) that a is at least a non-negative integer and consider the functions

$$(38)\qquad W(z) = CP(\delta)e^{-bz},$$

with $P(\delta) = (\delta-n-a+1)(\delta-n-a)\dots(\delta-2n-a+2)$ and constant C.

By (19)

$$(39)\qquad W = Cz^{2n+a-1}\frac{d^n}{dz^n}(z^{-n-a+1}e^{-bz}).$$

We claim that for any constant C, W is a solution of the differential equation (36). This can be proved directly, as in the second proof of Theorem 1, but an easier proof is by use of (18). On account of the homogeneity of (36), it is sufficient to verify the claim for C = 1. We obtain successively,

$$\delta(\delta+1-a-2n)W = \delta(\delta+1-a-2n)[(\delta-n-a+1)\dots(\delta-2n-a+2)e^{-bz}]$$

$$= (\delta-n-a+1)\dots(\delta-2n-a+2)(\delta-2n-a+1)\delta e^{-bz} =$$

$$(\delta-n-a+1)\dots(\delta-2n-a+1)(-zb)e^{-bz} = -bz(\delta-n-a+2)\dots(\delta-2n-a+2)e^{-bz}$$

$$= -bz(\delta-n-a+2)[(\delta-n-a+1)\dots(\delta-2n-a+2)e^{-bz}] = -bz(\delta-n-a+2)W,$$

as claimed. Here the first equality follows from (38), the second from the commutativity of the operators δ-c, the third from the definition of δ, the fourth from (18), and the last two are obvious.

One can easily verify that not all solutions W of (36), or, equivalently, of (33), are such that $e^{bz}w = p(z)$, a polynomial.

It follows that at most one of any two linearly independent solutions of (33) (or (36)) can be of this form. However, we do know one such solution, namely (38). Indeed, $e^{bz}W(z) = Ce^{bz}P(\delta)e^{-bz}$ is obviously a polynomial and, specifically, a constant multiple of $\theta_n(z;a,b)$ as already mentioned immediately after (33) and (36).

It follows, using (39) that, for an appropriate constant C, $\theta_n(z;a,b) = e^{bz}W(z) = Ce^{bz}z^{2n+a-1}\frac{d^n}{dz^n}(z^{-n-a+1}e^{-bz})$. In order to determine the constant C, we observe that the highest power of z furnished by the Leibniz form of the derivative $\frac{d^n}{dz^n}(z^{-n-a+1}e^{-bz})$ comes from the term $z^{-n-a+1}\frac{d^n}{dz^n}(e^{-bz})$ and is equal to $(-b)^n z^{-n-a+1}e^{-bz}$; hence, the right hand side equals

$$Ce^{bz}z^{2n+a-1}[(-b)^n z^{-n-a+1} + (\text{lower powers of } z)]e^{-bz} =$$

$Cz^n(-b)^n +$ (lower powers of z). The coefficient of z^n in $\theta_n(z;a,b)$ is 1, so that $C = (-1)^n b^{-n}$ and this finishes the proof of the Rodrigues-type formula

$$\theta_n(z;a,b) = (-1)^n b^{-n} e^{bz} z^{2n+a-1}\frac{d^n}{dz^n}(z^{-n-a+1}e^{-bz}). \tag{40}$$

For b = 2, in particular,

$$\theta_n(z;a) = (-1)^n 2^{-n} e^{2z} z^{2n+a-1}\frac{d^n}{dz^n}(z^{-n-a+1}e^{-2z})$$

and (see [17]), if also a = 2, then

$$\theta_n(z) = (-1)^n 2^{-n} e^{2z} z^{2n+1}\frac{d^n}{dz^n}(z^{-n-1}e^{-2z}).$$

7. It is clear that all values of $b \neq 0$ are admissible and lead to BP. For b = 0, however, (26) reduces to $z^2y'' + azy' - n(n+a-1)y = 0$. The solutions are of the form $y = z^r$, where the r are solutions of the equation $r(r-1)+ar-n(n+a-1) = 0$. These solutions are n and 1-a-n so that the general solution of (26) for b = 0 becomes $y = c_1z^n + c_2z^{1-a-n}$. For $n > a+1$, this is a polynomial only if $c_2 = 0$, when it reduces, essentially to z^n. This is a trivial case of no further interest here.

The parameter a may take arbitrary values. This is clear from the form of the coefficients $d_k^{(n)}$ (see (27)) and $f_k^{(n)}$, which remain well defined for all complex values of a. For some of the theory to be developed, however, the values a = 0, -1,-2,... lead to special difficulties and may have to be discussed separately. So, e.g. if a = 1-n, equation (26) becomes

$$\frac{y''}{y'} = \frac{n-1}{z} - \frac{b}{z^2}$$

and has the general solution $y = \frac{z^n}{n} + c_1z + c_2 + bz\log z$.

This is a polynomial only if $b = 0$, when it reduces to $y = c_1 z^n + c_2$, also of no further interest.

Whenever such cases arise, we shall tacitly or explicitly assume that a is not zero, or a negative integer.

CHAPTER 3

RECURRENCE RELATIONS

1. It is well known (see [1]) that the solutions $K_\nu(z)$ of (2.1) satisfy recurrence relations, such as

(1) $$K_{\nu-1}-K_{\nu+1} = -(2\nu/z)K_\nu,$$

(2) $$K'_\nu = -K_{\nu-1} - (\nu/z)K_\nu,$$

(3) $$K_{\nu-1}+K_{\nu+1} = -2K'_\nu,$$

(4) $$K'_\nu = -K_{\nu+1} + (\nu/z)K_\nu.$$

From (2.5) and $c_0 = (2n)!/2^n n!$ it follows that $K_{n+1/2}(z) = (\pi/2)^{1/2}e^{-z}z^{-n-1/2}\theta_n(z)$.

If we substitute this in (1) for $K_\nu(z)$, we obtain after routine simplifications (see [17])

(5) $$\theta_{n+1}(z) = (2n+1)\theta_n(z) + z^2\theta_{n-1}(z).$$

On account of $\theta_n(z) = e^z\phi_n(z)$, the same recurrence relation holds also for the functions $\phi_n(z)$:

(6) $$\phi_{n+1}(z) = (2n+1)\phi_n(z) + z^2\phi_{n-1}(z).$$

If we replace in (5) z by z^{-1}, multiply the result by z^{n+1} and recall that $z^n\theta_n(z^{-1}) = y_n(z)$, then we obtain (see [68]) the recurrence relation for the BP $y_n(z)$:

(7) $$y_{n+1}(z) = (2n+1)zy_n(z) + y_{n-1}(z).$$

The recurrence relations (5) and (7) allow us to compute successively the polynomials $\theta_n(z)$ and $y_n(z)$ with very little effort. We obtain as first few polynomials, starting from $\theta_0(z) = 1$ and $\theta_1(z) = 1+z$, the following:

$$\theta_2(z) = 3\theta_1(z) + z^2\theta_0(z) = z^2+3z+3;\ \theta_3(z) = 5\theta_2(z) + z^2\theta_1(z) =$$
$$5(z^2+3z+3) + z^2(z+1) = z^3+6z^2+15z+15, \text{ etc.}$$

Similarly, starting from $y_0(z) = 1$, $y_1(z) = 1+z$, we find $y_2(z) = 3zy_1(z) + y_0(z) = 3z^2+3z+1$, $y_3(z) = 5zy_2(z) + y_1(z) = 5z(3z^2+3z+1) + z+1 = 15z^3+15z^2+6z+1$, etc. These results illustrate the relations $z^ny_n(z^{-1}) = \theta_n(z)$ and $z^n\theta_n(z^{-1}) = y_n(z)$.

From $y_0(z) = \theta_0(z) = 1$, $y_1(z) = \theta_1(z) = z+1$, (5) and (7) also immediately follows

THEOREM 1. *The coefficients of all* BP *are positive, rational integers.*

If we proceed in the same way, by using also (see (2.5) solved for $\mu_\nu K_\nu(z)$), $\mu_\nu K_\nu' = -\nu z^{-\nu-1}e^{-z}\theta(z,\nu)-z^{-\nu}e^{-z}\theta(z,\nu) + z^{-\nu}e^{-z}\theta'(z,\nu)$, we obtain from (2) that θ_n, ϕ_n, and y_n satisfy

(8) $$\theta_n'(z) = \theta_n(z) - z\theta_{n-1}(z),$$

(9) $$\phi_n'(z) = -z\phi_{n-1}(z),$$

and

(10) $$z^2 y_n'(z) = (nz-1)y_n(z)+y_{n-1}(z),$$

respectively. For (8), see [17], for (10) see [68]. One may remark the particularly simple form of (9), which may be new.

Similarly, one obtains from (3) the following recurrence relations:

(11) $$2z\theta_n'(z) = (2z+2n+1)\theta_n(z) - (z^2\theta_{n-1}(z)+\theta_{n+1}(z)),$$

(12) $$2z\phi_n'(z) = -z^2\phi_{n-1}(z)+(2n+1)\phi_n(z)-\phi_{n+1}(z),$$

(13) $$\begin{aligned} 2z^2 y_n'(z) &= (y_{n-1}(z)-y_n(z))+(y_{n+1}(z)-y_n(z))-zy_n(z) \\ &= y_{n-1}(z)-(2+z)y_n(z)+y_{n+1}(z). \end{aligned}$$

Finally, (4) is a linear combination of (2) and (3). Either directly, as before, or by combining (8) and (11), or (9) and (12), or (10) and (13) we obtain (see [17] for (14)) the corresponding relations

(14) $$z\theta_n'(z) = (z+2n+1)\theta_n(z)-\theta_{n+1}(z),$$

(15) $$z\phi_n'(z) = (2n+1)\phi_n(z)-\phi_{n+1}(z),$$

(16) $$z^2 y_{n-1}'(z) = y_n(z)-(1+nz)y_{n-1}(z).$$

By combining above formulae, or by starting from any other among the numerous recurrence relations known in the theory of Bessel functions, many other recurrence relations for the polynomials $\theta_n(z)$ and $y_n(z)$, or for the functions $\phi_n(z)$ can be obtained.

For future use we observe that (7) (and similarly (5) and (6)), can be generalized to read

(7') $$y_n(z) = P_m(z)y_{n-m}(z) + Q_{m-1}(z)y_{n-m-1}(z),$$

where $P_m(z)$ and $Q_{m-1}(z)$ are polynomials in z of degrees m and m-1, respectively. Indeed, (7) is the instance m = 1 and the general statement follows by induction on m: Assuming (7') already verified, we replace $y_{n-m}(z)$ by use of (7) and obtain

$$\begin{aligned} y_n &= P_m(z)((2n-2m+1)zy_{n-m-1} + y_{n-m-2}) + Q_{m-1}(z)y_{n-m-1} \\ &= \{(2n-2m+1)zP_m(z) + Q_{m-1}(z)\}y_{n-m-1} + P_m(z)y_{n-m-2} \\ &= P_{m+1}(z)y_{n-m-1} + Q_m(z)y_{n-m-2}. \end{aligned}$$

This shows that (7') holds also for m+1, hence for all $m \leq n$ and, incidentally shows that $P_{m+1}(z) = (2n-2m+1)zP_m(z) + Q_{m-1}(z)$, $Q_m(z) = P_m(z)$.

Formulae (5), (6) and (7) permit us to extend the definitions of $\theta_n(z)$, $\phi_n(z)$, and $y_n(z)$ to negative values of n. For $\theta_n(z)$ and $y_n(z)$, this has already been done in (2.9) and (2.12) and one has to verify that the two extensions are consistent, for instance that (7) leads to $y_{-n} = y_{n-1}$.

From (7) it follows that $y_{n-1} = y_{n+1}-(2n+1)zy_n$; hence, $y_{-1} = y_1-zy_0 = 1+z-z = 1 = y_0$, in agreement with (2.12). Assuming that (7) remains consistent with (2.12) for all subscripts not exceeding n, by (7),

$$y_{-n-1} = y_{-n+1}-(-2n+1)zy_{-n} = y_{n-2}+(2n-1)zy_{n-1} = y_n$$

as we wanted to show. The consistency of (2.9) with (5) follows immediately from what precedes, and can also be checked directly by the same method.

2. The recurrence relation (7) permits us to give a new definition to the BP $y_n(x)$, as a determinant. Indeed, following [85] let us consider (7), with n replaced by n-1, as a homogeneous linear equation in the three "unknowns" y_n, y_{n-1}, y_{n-2}, namely $\frac{y_n}{2n-1} - zy_{n-1} - \frac{y_{n-2}}{2n-1} = 0$. Next, we replace n successively by n-1, n-2,..., 3,2, and 1. For n = 3 the equation reads $\frac{y_3}{5} - zy_2 - \frac{y_1}{5} = 0$. For n = 2 we write it (inhomogeneously) in the form $\frac{y_2}{3} - zy_1 = \frac{1}{3}$ $(= \frac{y_0}{3})$. Finally, for n = 1 (the last one), it is $y_1 = z+1$ $(= zy_0 + y_{-1})$. We solve this system of n linear equations in $y_n, y_{n-1},\ldots,y_1$ by Cramer's rule. The determinant of the system is

$$D = \begin{vmatrix} \frac{1}{2n-1} & -z & -\frac{1}{2n-1} & 0 & 0 & \dots & & 0 \\ 0 & \frac{1}{2n-3} & -z & -\frac{1}{2n-3} & 0 & \dots & & 0 \\ \vdots & \vdots & \vdots & \vdots & \vdots & & & \vdots \\ 0 & 0 & 0 & 0 & 0 & \dots & \frac{1}{3} & -z \\ 0 & 0 & 0 & 0 & 0 & \dots & 0 & 1 \end{vmatrix}$$

and $y_n D = D_n$, where D_n is obtained by replacing the first column of D by the column vector of second members of the equations. The entries of this vector are, as seen, all zeros, except for the last two, which are 1/3 (from n = 2) and z+1 (from n = 1), respectively. It is clear that $D = \{(2n-1)(2n-3)\dots 3.1\}^{-1}$, because D is triangular; hence it is equal to the product of its diagonal elements. As for

$$D_n = \begin{vmatrix} 0 & -z & -\frac{1}{2n-1} & 0 & \dots & 0 & 0 \\ 0 & \frac{1}{2n-3} & -z & -\frac{1}{2n-3} & \dots & 0 & 0 \\ 0 & 0 & \frac{1}{2n-5} & -z & \dots & 0 & 0 \\ \vdots & \vdots & \vdots & \vdots & & \vdots & \vdots \\ \frac{1}{3} & 0 & 0 & 0 & \dots & \frac{1}{3} & -z \\ z+1 & 0 & 0 & 0 & \dots & 0 & 1 \end{vmatrix}$$

we shift the first column to last place, taking into account the sign change $(-1)^{n-1}$.

Next, we change the signs of the entries in the last column, so that $D_n = (-1)^n M_n$, where

$$M_n = \begin{vmatrix} -z & -\frac{1}{2n-1} & 0 & \dots & 0 & 0 & 0 \\ \frac{1}{2n-3} & -z & -\frac{1}{2n-3} & \dots & 0 & 0 & 0 \\ 0 & \frac{1}{2n-5} & -z & \dots & 0 & 0 & 0 \\ \vdots & \vdots & \vdots & & \vdots & \vdots & \vdots \\ 0 & 0 & 0 & \dots & \frac{1}{3} & -z & -\frac{1}{3} \\ 0 & 0 & 0 & \dots & 0 & 1 & -z-1 \end{vmatrix}$$

Finally, $y_n(z) = D_n/D = (-1)^n.1.3.5 \ldots (2n-1)M_n$.

The corresponding results for the generalized BP $y_n(z;a,b)$ do not appear equally useful, because they are rather complicated and lack symmetry (see, however [74]).

For $\theta_n(z)$ on the other hand, we may start from (5) (with n replaced by n-1) instead of (7) and obtain in exactly the same way that

$D\theta_n = D_n$, where

$$D = \begin{vmatrix} 1 & -(2n-1) & -z^2 & 0 & \ldots & 0 & 0 & 0 \\ 0 & 1 & -(2n-3) & -z^2 & \ldots & 0 & 0 & 0 \\ \vdots & \vdots & \vdots & \vdots & & \vdots & \vdots & \vdots \\ 0 & 0 & 0 & 0 & \ldots & 1 & -5 & -z^2 \\ 0 & 0 & 0 & 0 & \ldots & 0 & 1 & -3 \\ 0 & 0 & 0 & 0 & \ldots & 0 & 0 & 1 \end{vmatrix} = 1$$

and

$$D_n = (-1)^n \begin{vmatrix} -(2n-1) & -z^2 & 0 & \ldots & 0 & 0 & 0 \\ 1 & -(2n-3) & -z^2 & \ldots & 0 & 0 & 0 \\ 0 & 1 & -(2n-5) & \ldots & 0 & 0 & 0 \\ \vdots & \vdots & \vdots & & \vdots & \vdots & \vdots \\ 0 & 0 & 0 & \ldots & -5 & -z^2 & 0 \\ 0 & 0 & 0 & \ldots & 1 & -3 & -z^2 \\ 0 & 0 & 0 & \ldots & 0 & 1 & -(1+z) \end{vmatrix},$$

with $\theta_n(z) = D_n$.

3. From the recurrence relations for the BP, we obtain recurrence relations for their coefficients. We denote as before (see (2.8)) by $a_m^{(n)}$ the coefficient of z^m in $y_n(z)$, normalized by $y_n(0) = 1$.

By substituting (2.10), i.e. $y_n(z) = \sum_{m=0}^{n} a_m^{(n)} z^m$ into (7) we obtain

$$(17) \qquad a_m^{(n)} = \begin{cases} 1 \quad \text{for } m = 0, & \\ (2n-1)a_{m-1}^{(n-1)} + a_m^{(n-2)} & \text{for } 1 \le m \le n-2, \\ (2n-1)a_{m-1}^{(n-1)} & \text{for } m = n-1 \text{ and } m = n. \end{cases}$$

By substituting (2.10) into (10) we obtain

$$(18)\qquad a_m^{(n)} = \begin{cases} a_0^{(n-1)} = 1 & \text{for } m = 0, \\ (n-m+1)a_{m-1}^{(n)} + a_m^{(n-1)} & \text{for } 1 \le m \le n-1, \\ a_n^{(n)} & \text{for } m = n. \end{cases}$$

Finally, if we substitute (2.10) into (16) we obtain

$$(19)\qquad a_m^{(n)} = \begin{cases} 1 & \text{for } m = 0\ , \\ (m+n-1)a_{m-1}^{(n-1)} + a_m^{(n-1)} & \text{for } 1 \le m \le n-1\ , \\ (2n-1)a_{n-1}^{(n-1)} & \text{for } m = n. \end{cases}$$

In each set of formulae (17), (18), and (19), the central formula holds for all cases, if we remember that $a_0^{(n)} = 1$ and $a_m^{(n)} = 0$ for $m > n$, or $m < 0$.

4. Recurrence relations similar to (5) and (7) hold for the generalized BP and for $\phi_n(z,a)$. However, their usefulness seems to be limited and the proofs are routine.

For these reasons it should suffice to only list a few of them here (some of these appear in [68]):

$$(20)\qquad (n+a-1)(2n+a-2)y_{n+1}(z,a) = [(2n+a)(n-1+a/2)z+a-2](2n+a-1)y_n(z,a) + n(2n+a)y_{n-1}(z,a),$$

$$(21)\qquad z^2(2n+a-2)y_n'(z,a) = [n(2n+a-2)z-2n]y_n(z,a)+2ny_{n-1}(z,a)\ ,$$

$$(22)\qquad z^2(2n+a-2)y_{n-1}'(z,a) = 2(n+a-2)y_n(z,a) - \{(n+a-2)(2n+a-2)z+2n\}y_{n-1}(z,a)\ ,$$

$$(23)\qquad (n+a-1)(2n+a-2)\theta_{n+1}(z,a) = [(2n+a)(n-1+a/2)+(a-2)z](2n+a-1)\theta_n(z,a) + n(2n+a)z^2\theta_{n-1}(z,a),$$

$$(24)\qquad (1+(a-2)/2n)\theta_n'(z,a) = \theta_n(z,a) - z\theta_{n-1}(z,a).$$

Recursion (23) holds unchanged if we replace $\theta_n(z,a)$, by $\phi_n(z,a) = \theta_n(z,a)e^{-z}$; however, (24) becomes

$$(25)\qquad (2n+a-2)\phi_n'(z,a) = -(a-2)\phi_n(z,a)-2nz\phi_{n-1}(z,a).$$

One sees from (25), that the surprizing simplicity of (9) was, in some sense, rather accidental.

5. The reader may have been surprized that in Chapter 2 nothing was said concerning generalized BP for negative subscripts. This was no simple omission. The recurrence relation (20) permits us to define $y_{-n-1}(z;a,b)$ recursively, if we know y_{-n} and y_{1-n}. The first step, however, the determination of $y_{-1}(z;a,b)$ fails because in (20) the factor $n(2n+a)$ of y_{n-1} vanishes for $n = 0$. This is not the case when $a = 2$; in that case n (in fact $2n(n+1)$) is a factor common to all terms and is cancelled. Also the difficulty with $y_{-n}(z;a,b)$ for arbitrary real a is not an accident. Indeed, let us consider the differential equation (2.26) with $n = -1$,

$$z^2y'' + (az+b)y' + (a-2)y = 0.$$

We easily find that it has a single solution regular at $z = 0$ and normalized by $y(0) = 1$, given by

$$(26) \qquad y(z)(=y_{-1}(z;a,b)) = \sum_{m=0}^{\infty} c_m z^m, \text{ with } c_m = (-1)^m \frac{(m+a-3)^{(m)}}{b^m}.$$

This solution reduces to a polynomial if and only if a equals an integer not in excess of 2. For $a = 2$, $y(z)$ (i.e., $y_{-1}(z;2,b)$) reduces to the constant 1 ($= y_0(z;2,b)$), as we know from (2.12). The other case of interest is $a = 1$, when $y_{-1}(z;1,b) = 1+z/b = y_1(z;1,b)$ and, by (20) with $a = 1$, $y_{-n}(z;1,b)$ is a polynomial of exact degree n. However, the theory for $a = 1$ has never been developed much further, the suggestion of Krall and Frink ([68], bottom of page 109) notwithstanding. See, however, [99] and [100].

For the other cases when $y_{-1}(z;a,b)$ is a polynomial, that is, for $a = 0$, or a equal to a negative integer, $y_{-1}(z;a,b)$ is given by (26) and once we know $y_{-1}(z;a,b)$ all other $y_{-n}(z;a,b)$ can then be obtained recursively from (20), and are polynomials of degree $n+1-a$.

Unfortunately, some parts of the theory break down, or become complicated precisely in these cases, when $a = 0, -1, -2, \ldots$ and so we shall not pursue this matter further.

CHAPTER 4

MOMENTS AND ORTHOGONALITY ON THE UNIT CIRCLE

1. The theory of moments has a long history. Its modern phase may be said to have started with Stieltjes [**54**] and it was considerably advanced by the work of Tchebycheff [**57**] and Hamburger [**26**], [**27**]. It is intimately connected with the idea of orthogonality.

In the classical setting, one starts with a function $w(x)$, real and non-negative over an interval $a \leqq x \leqq b$ of the real axis, and considers the "moments" $m_n = \int_a^b w(x)x^n dx$. Set

$$\Delta_n = \Delta_n(w) = \begin{vmatrix} m_0 & m_1 & \cdots & m_{n-1} \\ m_1 & m_2 & \cdots & m_n \\ \vdots & \vdots & & \vdots \\ m_{n-1} & m_n & \cdots & m_{2n-2} \end{vmatrix}, \quad \Delta_0 = 1$$

and let us assume that $\Delta_n \neq 0$ for all $n = 1,2,\ldots$.

Next, we recall that the polynomials with real coefficients form a vector space S over the reals. We define an inner product on S by setting

$$(f,g) = \int_a^b w(x)f(x)g(x)dx.$$

Clearly, $(f,g) = (g,f)$. If $(f,g) = 0$, we say that f and g are orthogonal to each other.

One may take as a basis for S the powers of x, $x^0 = 1$, x, $x^2, \ldots$ which are linearly independent.

By use of the Schmidt method (see [**38**], p. 152]) - sometimes called the Gram-Schmidt procedure - we can find a finite sequence of linear combinations of those powers, i.e., of polynomials $p_n(x)$ of successive degrees $0,1,2,\ldots$, such that each one is orthogonal to all preceding ones. If $p_n(x) = \sum_{m=0}^{n} c_m x^{n-m}$ then it is, of course, sufficient to insure that $(x^k, p_n(x)) = 0$ for $k = 0,1,\ldots,n-1$. If we substitute here $\sum_{m=0}^{n} c_m x^{n-m}$ for $p_n(x)$ and use the definition and linearity of the inner product, we obtain n homogeneous equations in the n+1 coefficients $c_m (m = 0,1,\ldots,n)$ of $p_n(x)$. These have a unique solution, up to an arbitrary multiplicative factor, if and only if their determinant is different from zero. One immediately verifies that this determinant is precisely Δ_n and, by assumption, $\Delta_n \neq 0$. We may take advantage of the mentioned arbitrary constant factor of the c_m's, in order to obtain also $(p_n,p_n) = 1$. A set of polynomials that satisfies these conditions, i.e. $(p_n, p_m) = \delta_{nm}$, is called orthonormal. One may show (see [56], 2.2.6) that

$$p_n(x) = (\Delta_{n-1} \cdot \Delta_n)^{-1/2} \begin{vmatrix} m_0 & m_1 & \dots & m_n \\ m_1 & m_2 & \dots & m_{n+1} \\ \vdots & \vdots & & \vdots \\ m_{n-1} & m_n & \dots & m_{2n-1} \\ 1 & x & \dots & x^n \end{vmatrix}$$

For more details see [56].

This situation is usually generalized as follows (see, e.g. [56], [69], or [63]): One replaces the interval [a,b] of the real axis by an arc (infinite, or finite, open or closed) of a curve C in the complex plane and defines the inner product by $(f,g) = \int_C w(z)f(z)\bar{g}(z)|dz|$, where $w(z)$ is a (generally non-holomorphic) function of the complex variable z, real and non-negative on C. One observes that instead of the commutativity $(f,g) = (g,f)$, the inner product now satisfies the equation $(f,g) = \overline{(g,f)}$. Much of the classical theory goes through in this broader setting. For the case in point, however, a different kind of orthogonality is of interest. It occurs already in [68] and has been studied in detail by John W. Jayne [63] who calls it the Bessel alternative. Still more general points of view are possible. R.D. Morton and A.M. Krall [80] start from a weight "function" that is, in fact a distribution. In the case of the classical orthonormal polynomials (Jacobi, Laguerre, Hermite), use of the Fourier transform permits to recover the usual weight functions from the distributional ones. This approach is not entirely successful in the case of the BP. A recent paper of A.M. Krall [66] however, completes this work.

In spite of the great interest of this problem, it will not be discussed here further. Indeed, precisely because of its breath, it encompasses much that is not directly related to BP and the interested reader is directed to the original papers.

We shall restrict ourselves to problems of moments directly pertinent to BP and, indeed, many of the results of this chapter can be found in [68] and in [17], although not always in the same generality.

For arbitrary real or complex $a \neq 0, -1, -2, \dots$ and $z \neq 0$, set $\rho = \rho(z;a,b) = \sum_{n=0}^{\infty} \frac{\Gamma(a)}{\Gamma(a+n-1)} (-\frac{b}{z})^n$ and let $p_n(z) = \sum_{m=0}^{n} c_m z^m$ be an arbitrary polynomial of degree n. The k-th moment of $p_n(z)$ on the unit circle, with weight function ρ is defined by

$$M_k(p_n;\rho) = \frac{1}{2\pi i} \int_{|z|=1} z^k p_n(z) \left(\sum_{r=0}^{\infty} \frac{\Gamma(a)}{\Gamma(a+r-1)} (-b/z)^r \right) dz.$$

Here we may substitute the explicit form of $p_n(z)$ and interchange summation and integration because the sum converges essentially like $\sum_{n=0}^{\infty} \frac{|b|^n}{n!}$, $p_n(z)$ contains only a finite number of terms, and the path of integration is finite. Most terms vanish and we obtain

$$M_k(p_n,\rho) = \sum_{k+m-r=-1} c_m(\Gamma(a)/\Gamma(a+r-1))(-1)^r b^r = \sum_{r=k+1}^{n+k+1} (-1)^r b^r c_{r-k-1}\Gamma(a)/\Gamma(a+r-1).$$

We are interested, in particular, in the case when $p_n(x) = y_n(z;a,b) = \sum_{m=0}^{n} f_m^{(n)} z^m$, where (see (2.28)) $f_m^{(n)} = \binom{n}{m} b^{-m}(n+m+a-2)(n+m+a-3)\dots(n+a-1)$. The sum may then be written successively, as follows:

$$\sum_{r=k+1}^{n+k+1} (-1)^r b^r \frac{n!(n+r-k+a-3)\dots(n+a-1)\Gamma(a)}{(r-k-1)!(n-r+k+1)!b^{r-k-1}\Gamma(a+r-1)} =$$

$$\frac{(-b)^{k+1}}{(k+n+1)!} \sum_{v=r-k-1=0}^{n} (-1)^v \frac{n!(n+v+a-2)\dots(n+a-1)\cdot(n-v+1)\dots(n+k+1)}{v!\; a(a+1)\dots(a+v+k-1)} =$$

$$\frac{(-b)^{k+1}}{(k+n+1)!} \sum_{v=0}^{n} (-1)^v \frac{n!}{v!} \frac{(n+k+1)!}{(n-v)!} \frac{(n+v+a-2)\dots(n+a-1)}{(k+v+a-1)\dots(a+1)a} =$$

$$(-b)^{k+1} \sum_{v=0}^{n} (-1)^v \binom{n}{v} \frac{(n+v+a-2)\dots(n+a-1)}{(k+v+a-1)\dots(a+1)a},$$

so that

(1) $$M_k(y_n,\rho) =$$

$$\frac{(-b)^{k+1}}{(k+n+a-1)\dots(a+1)a} \sum_{v=0}^{n} (-1)^v \binom{n}{v} (n+v+a-2)\dots(n+a)(n+a-1)\cdot(k+v+a)(k+v+a+1)\dots(k+n+a-1).$$

We claim that the sum is independent of a and is equal to the constant $k(k-1)\dots(k-n+1)$. If we substitute this value for the sum we obtain

THEOREM 1. *Let* $y_n = y_n(z;a,b)$ *and* $\rho = \rho(z;a,b)$; *then*

$$M_k(y_n;\rho) = (-b)^{k+1} \frac{\Gamma(a)\Gamma(k+1)}{\Gamma(k+a+n)\Gamma(k+1-n)}.$$

Our claim concerning the sum may be well-known (see e.g. [68], p. 113); however, as no reference appears to be readily available, a proof of the claim follows. It is based on two lemmas, the first of which is indeed well-known.

LEMMA 1. (see (3.4) in [20]). $\sum_{v=0}^{t} \binom{n}{v}\binom{k-n}{t-v} = \binom{k}{t}$.

<u>Proof.</u> $\sum_{t=0}^{k} \binom{k}{t} x^t = (x+1)^k = (x+1)^n (x+1)^{k-n} = \sum_{v=0}^{n} \binom{n}{v} x^v \sum_{w=0}^{k-n} \binom{k-n}{w} x^w$

$$= \sum_{t=0}^{k} x^t \sum_{\substack{v+w=t \\ 0 \le v \le n \\ 0 \le w \le k-n}} \binom{n}{v}\binom{k-n}{w} = \sum_{t=0}^{k} x^t \sum_{v=0}^{n} \binom{n}{v}\binom{k-n}{t-v}$$

and the lemma follows from the comparison of the coefficients of x^t in the first and last member.

<u>LEMMA 2.</u> *The function*

$$f(x) = \sum_{v=0}^{n} (-1)^v \binom{v}{n} \frac{\Gamma(k+n+x+1)}{\Gamma(n+x)} \frac{1}{(k+v+x)(k-1+v+x)\dots(n+v+x)} \tag{2}$$

is independent of x, *and has the value* $k(k-1)\dots(k-n+1)$; *in particular, for* $k < n$, *one has* $f(x) = 0$.

<u>Proof.</u> For $k \geq n$ (2) is equivalent to

$$f(x) = \sum_{v=0}^{n} (-1)^v \binom{n}{v} \frac{(k+n+x)(k+n+x-1)\dots(k+v+x)\dots(n+v+x)\dots(n+x)}{(k+v+x)\dots(n+v+x)} . \tag{3}$$

$f(x)$ is a rational function and could have poles at most for $x = -m$, $n \leq m \leq k+n$. In fact, as follows from (3), all these poles cancel and $f(x)$ is a polynomial of degree at most n. We shall show that $\lim_{x \to -m} f(x)$ exists for all these k+1 values of m in the given range and that all these limits equal $k!/(k-n)!$, independently of m. It then follows that the polynomial $f(x)$ of degree $n < k+1$ equals $k!/(k-n)!$ identically, as claimed.

To find the limit, set $m = n+t$, $0 \leq t \leq n$, $t \in Z$, $x = -n-s$. Then, by (2),

$$\lim_{s \to t} f(-n-s) = \lim_{s \to t} \frac{\Gamma(k-s+1)}{(t-s)(t-1-s)\dots(-s)\Gamma(-s)} \sum_{v=0}^{n} (-1)^v \binom{n}{v} \frac{(t-s)\dots(-s)}{(k+v-n-s)\dots(v-s)}$$

$$= \Gamma(k-t+1) \lim_{s \to t} \sum_{v=0}^{t} (-1)^v \binom{n}{v} \frac{t(t-1)\dots(t-v+1)}{(k+v-n-t)!} ,$$

because for $t < v \leq n$, $-s < t-s < v-s$ and the vanishing factor of the numerator is not cancelled by a corresponding factor of the denominator. It follows that

$$\lim_{s \to t} f(-n-s) = \Gamma(k-t+1) \sum_{v=0}^{t} (-1)^v \binom{n}{v} \frac{t!}{(t-v)!(k+v-n-t)!} = \frac{(k-t)!t!}{(k-n)!} \sum_{v=0}^{t} (-1)^r \binom{n}{v}\binom{k-n}{t-v} .$$

By Lemma 1, the last sum equals $\binom{k}{t}$, so that indeed

$$\lim_{x \to -m} f(x) = \frac{(k-t)!t!}{(k-n)!} \frac{k!}{t!(k-t)!} = \frac{k!}{(k-n)!} \text{, as claimed.}$$

Proof of Theorem 1. In the sum in (1) set a-1 = x. The sum becomes

$$\sum_{v=0}^{n} (-1)^v \binom{n}{v} (n+v+x-1)\ldots(n+x)\cdot(k+v+x-1)\ldots(k+n+x)$$

$$= \sum_{v=0}^{n} (-1)^v \binom{n}{v} \frac{(k+n+x)(k+n+x-1)\ldots(k+v+x)\ldots(n+v+x)\ldots(n+x)}{(k+v+x)\ldots(n+v+x)} = \frac{k!}{(k-n)!}$$

by Lemma 2. If we substitute this value in (1), the theorem follows.

Concerning the range of validity of Theorem 1, one observes that the right hand side has a finite value for all $a \neq -m$, m an integer of the interval $0 \leq m \leq k+n-1$. If a approaches one of the excluded values, then $M_k(y_n(z;a,b),\rho) \to \infty$.

2. From Theorem 1 follow several corollaries.

COROLLARY 1. (see (57) in [68]; also Lemma 16.1 in [80]). *For* $0 \leq k < n$, $M_k(y_n(z;a,b);\rho) = 0$.

Formally, the corollary follows immediately from Theorem 1. It can also be verified directly. Indeed, now we obtain instead of (1):

$$M_k(y_n,\rho) = (-b)^{k+1} \sum_{v=0}^{n} (-1)^v \binom{n}{v} \frac{(n+v+a-2)\ldots(n+a-1)}{(k+v+a-1)\ldots(a+1)a}$$

and the sum, again with a-1 = x, equals

$$\frac{1}{(x+1)\ldots(x+n-1)} \sum_{v=0}^{n} (-1)^v \binom{n}{v} (n+v+x-1)(n+v+x-2)\ldots(k+v+x+1).$$

This sum vanishes for k < n. Indeed, let $F(y) = \sum_{v=0}^{n} (-1)^v \binom{n}{v} y^{n+v+x-1}$; then

$$\frac{\partial^{n-k-1}F}{\partial y^{n-k-1}} = \sum_{v=0}^{n} (-1)^v \binom{n}{v} (n+v+x-1)\ldots(k+v+x+1)y^{k+v+x}.$$

The sum to be evaluated is the last one for y = 1. On the other hand,

$$F(y) = y^{n+x-1} \sum_{v=0}^{n} (-1)^n \binom{n}{v} y^v = y^{n+x-1}(1-y)^n.$$

Hence, by Leibniz' rule, the (n-k-1)-th derivative is of the form

$\sum_{s+t=n-k-1} \alpha_s y^{n+x-1-s}(1-y)^{n-t}$ and vanishes for y = 1, because $n-t \geq n-(n-k-1) = k+1 \geq 1$.

By taking for a a natural integer and, in particular a = 2, we obtain from Theorem 1 the following corollaries:

COROLLARY 2. (see [68]). *For integral* a > 0,

$$M_k(y_n(z;a,b);\rho) = (-b)^{k+1} \frac{(a-1)!k!}{(k+n+a-1)!(k-n)!} .$$

COROLLARY 3. (see [68]). *For* $a = 2$, $M_k(y_n(z;2,b);\rho) = (-b)^{k+1} \frac{k!}{(k+n+1)!(k-n)!}$.

REMARK. For $a = 2$, $\rho(z;2,b) = \sum_{n=0}^{\infty} \frac{(-1)^n (b/z)^n}{n!} = e^{-b/z}$ so, that Corollary 3 may be rephrased as

COROLLARY 3'. (see [68]). *For* $k \geq n$

$$\frac{1}{2\pi i} \int_{|z|=1} z^k y_n(z;2,b) e^{-b/z} dz = (-b)^{k+1} \frac{k!}{(k+n+1)!(k-n)!}$$; *the integral vanishes for* $k < n$.

For $a = b = 2$, in particular, we obtain

COROLLARY 4. (see [68]).

$$M_k(y_n(z),\rho) = M_k(y_n, e^{-2/z}) = \frac{1}{2\pi i} \int_{|z|=1} z^k y_n(z) e^{-2/z} dz = \frac{(-2)^{k+1} k!}{(k+n+1)!(k-n)!} \quad \textit{for } k \geq n$$

and vanishes for $k < n$.

COROLLARY 5. (see [17]). *If* $k \neq n$, *then*

$$\frac{1}{2\pi i} \int_{|z|=1} y_k(z;a,b) y_n(z;a,b) \rho(z;a,b) dz = 0.$$

Proof. Let $k < n$; then $y_k(z;a,b) = \sum_{m=0}^{k} f_m^{(k)} z^m$, with $0 \leq m < n$; hence, the result follows from Corollary 1.

COROLLARY 6. (see [68], [17]).

$$\frac{1}{2\pi i} \int_{|z|=1} y_n^2(z;a,b) \rho(z;a,b) dz = (-1)^{n+1} \frac{bn!}{2n+a-1} \frac{\Gamma(a)}{\Gamma(n+a-1)} .$$

Proof. By Corollary 5, $\frac{1}{2\pi i} \int_{|z|=1} y_n^2(z;a,b) \rho(z;a,b) dz$

$$= \frac{1}{2\pi i} \int_{|z|=1} f_n^{(n)} z^n y_n(z;a,b) \rho(z;a,b) dz = f_n^{(n)} M_n(y_n(z;a,b),\rho)$$

and the result follows from Theorem 1 and (2.28).

By setting $a = b = 2$, one obtains from Corollary 6 also

COROLLARY 7. (see [68]) $\frac{1}{2\pi i} \int_{|z|=1} y_n^2(z) e^{-2/z} dz = (-1)^{n+1} \frac{2}{2n+1}$.

3. If we replace $y_n(z;a,b)$ by $z^n \theta_n(z^{-1};a,b)$ in Theorem 1 and its corollaries, we obtain the following results.

THEOREM 2. *Let* $\rho_1(z;a,b) = z^{-2}\sum_{r=0}^{\infty} \frac{\Gamma(a)}{\Gamma(a+r-1)}(-bz)^r$ *and set*

$$\tilde{M}_k(\theta_n;\rho_1) = \tilde{M}_k(\theta_n(z;a,b);\rho_1(z;a,b)) = \frac{1}{2\pi i}\int_{|z|=1} z^{-k}(z^{-n}\theta_n(z;a,b))\rho_1(z;a,b)dz;$$

then

$$\tilde{M}_k(\theta_n;\rho_1) = (-b)^{k+1}\frac{\Gamma(a)\Gamma(k+1)}{\Gamma(k+a+n)\Gamma(k+1-n)}.$$

COROLLARY 8. *For* $k < n$, $\tilde{M}_k(\theta_n,\rho_1) = 0$.

COROLLARY 9. *For integral* $a > 0$, $\tilde{M}_k(\theta_n;\rho_1) = (-1)^{k+1}\frac{b^{k+1}(a-1)!k!}{(k+n+a-1)!(k-n)!}$.

For $a = 2$, $\rho_1(z;2,b) = z^{-2}\sum_{r=0}^{\infty}\frac{(-bz)^r}{r!} = z^{-2}e^{-bz}$ and we obtain

COROLLARY 10. *For* $a = 2$, $\tilde{M}_k(\theta_n,z^{-2}e^{-bz}) = (-b)^{k+1}\frac{k!}{(k+n+1)!(k-n)!}$.

In particular, for $a = b = 2$, one obtains

COROLLARY 11. *For* $a = b = 2$, $\tilde{M}_k(\theta_n,z^{-2}e^{-2z}) = (-2)^{k+1}\frac{k!}{(k+n+1)!(k-n)!}$.

COROLLARY 12. *If* $k \neq n$, *then*

$$\frac{1}{2\pi i}\int_{|z|=1}(z^{-k}\theta_k(z;a,b))(z^{-n}\theta_n(z;a,b))\rho_1(z;a,b)dz = 0.$$

COROLLARY 13. *If* $k = n$, *then*

$$\frac{1}{2\pi i}\int_{|z|=1}(z^{-n}\theta_n(z;a,b))^2\rho_1(z;a,b)dz = (-1)^{n+1}b\frac{n!}{2n+a-1}\frac{\Gamma(a)}{\Gamma(n+a-1)}.$$

In particular, for $a = b = 2$, the following Corollary holds:

COROLLARY 14. $\frac{1}{2\pi i}\int_{|z|=1}(z^{-n}\theta_n(z))^2 z^{-2}e^{-2z}dz = (-1)^{n+1}\frac{2}{2n+1}$.

4. In the case of positive integral a, Corollaries 12 and 13 may be written in a somewhat different way.

Let us observe that in this case

$$\rho_1(z;a,b) = z^{-2}\sum_{r=0}^{\infty}\frac{(-bz)^r}{a(a+1)\dots(a+r-2)} = (a-1)!z^{-2}\sum_{r=0}^{\infty}\frac{(-bz)^r}{(a+r-2)!}$$

$$= b^2(a-1)!(-bz)^{-a}\sum_{\nu=a-2}^{\infty}\frac{(-bz)^\nu}{\nu!}$$

$$= b^2(a-1)!(-bz)^{-a}\left(e^{-bz}-\sum_{r=0}^{a-3}\frac{(-bz)^r}{r!}\right).$$

Hence, $\int_{|z|=1} (z^{-k}\theta_k(z;a,b))(z^{-n}\theta_n(z;a,b))\rho_1(z;a,b)dz =$

$$(-1)^a b^{2-a}(a-1)! \int_{|z|=1} z^{-a-k-n}\theta_k(z;a,b)\theta_n(z;a,b)(e^{-bz} - \sum_{r=0}^{a-3} \frac{(-bz)^r}{r!})\, dz$$

$$= (-b)^{2-a}(a-1)! \int_{|z|=1} z^{-a-k-n}\theta_k(z;a,b)\theta_n(z;a,b)e^{-bz}dz.$$

Indeed, for $0 \le r \le a-3$, $z^{-a-k-n+r}$ has an exponent $t = -a-k-n+r \le -k-n-3$; hence, all neglected integrals are of the form $\int_{|z|=1} z^{-t}\theta_n(z;a,b)\theta_k(z;a,b)dz$ and vanish, because the highest power of z that $\theta_n\theta_k$ can contribute is z^{k+n}.

It follows that for $a \in \mathbb{Z}^+$ the integrals in Corollaries 12 and 13 may be written as

$$(-b)^{2-a}(a-1)! \frac{1}{2\pi i} \int_{|z|=1} z^{-a-k-n}\theta_k(z;a,b)\theta_n(z;a,b)e^{-bz}dz = \delta_{nk}(-1)^{n+1} b \frac{n!(a-1)!}{(2n+a-1)(n+a-2)!}.$$

As a consequence the following Corollary holds:

<u>COROLLARY 15</u>. (see [17]). *For* k = n,

$$\frac{1}{2\pi i} \int_{|z|=1} z^{-a-k-n}\theta_k(z;a,b)\theta_n(z;a,b)e^{-bz}dz = (-1)^{n+a+1} b^{a-1} \frac{n!}{(2n+a-1)\cdot(n+a-2)!};$$

for $k \neq n$ *the integral vanishes.*

If we replace here z by z^{-1} and write $y_n(z;a,b)$ for $z^n\theta_n(z^{-1};a,b)$, we obtain

<u>COROLLARY 16.</u>

$$\frac{1}{2\pi i} \int_{|z|=1} z^{a-2} y_n(z;a,b)y_k(z;a,b)e^{-b/z}dz = \delta_{kn}(-1)^{n+a+1} b^{a-1} \frac{n!}{(2n+a-1)(n+a-2)!}.$$

For a = b = 2, Corollaries 15 and 16 reduce to Corollaries 14 and 7, respectively.

Finally, if we recall that, by (2.29), $\phi_n(z;a,b) = e^{-bz/2}\theta_n(z;a,b)$, we obtain from Corollaries 12, 14 and 15 that, for $a \in \mathbb{Z}^+$

$$\frac{1}{2\pi i} \int_{|z|=1} z^{-a}(z^{-k}\phi_k(z;a,b))(z^{-n}\phi_n(z;a,b))dz = (-1)^{n+a+1} b^{a-1} \frac{n!}{(2n+a-1)\cdot(n+a-2)!}$$

if k = n, = 0 if $k \neq n$.

In particular, $\frac{1}{2\pi i} \int_{|z|=1} z^{-2}(z^{-n}\phi_n(z))^2 dz = (-1)^{n+1} \frac{2}{2n+1}$.

Note added August 30, 1978. In the present chapter we discussed only the orthogonality on the unit circle and the corresponding moments. In fact, it was known from general considerations, going back to the work of Stieltjes, that a function $\phi(x)$, of bounded variation, ought to exist, such that $\int_{-\infty}^{\infty} y_n(x)y_m(x)d\psi(x) = 0$ should hold for all integers $m \neq n$, with the integral taken along the real axis. Unfortunately, no such function $\psi(x)$ was known.

At the moment of shipping the present typescript to the publisher, the author receives a (handwritten) manuscript by A.M. Krall, with the following important results.

Define the function $\psi(x)$ *by*

$$\psi(\beta)-\psi(\alpha) = \lim_{\varepsilon \to 0} \frac{-1}{\pi} \int_{\alpha}^{\beta} \exp\left(\frac{-2x}{x^2+\varepsilon^2}\right)\sin\left(\frac{2\varepsilon}{x^2+\varepsilon^2}\right)dx;$$

then the following holds:

1. $\psi(x)$ *is of bounded variation.*
2. *The support of* $d\psi(x)$ *is the origin.*
3. *The moments* $\mu_n = \int_{-\infty}^{\infty} x^n d\psi(x) = (-2)^{n+1}/(n+1)!$ for $n = 0,1,2,\ldots$
4. $\int_{-\infty}^{\infty} y_n(x)y_m(x)d\psi(x) = (-1)^{n+1}2\delta_{mn}/(2n+1)$, where δ_{mn} *stands for the Kronecker delta.*
5. *For* $(a,b) \neq (2,2)$, *these results generalize as follows:*

   Let $z = x+i\varepsilon$ *and set*

$$\psi(\beta)-\psi(\alpha) = \lim_{\varepsilon \to 0} \frac{1}{\pi} \int_{\alpha}^{\beta} I_m\{(-b/z)\,{}_1F_1\left[{1 \atop a};\ -b/z\right]\}dx;$$

   then $\psi(x)$ *is of bounded variation,* $\mu_n = (-b)^{n+1}/a(a+1)\ldots(a+n-1)$, *and* $\int_{-\infty}^{\infty} y_n(x;a,b)y_m(x;a,b)d\psi(x) = (-1)^{n+1}n!b\,\delta_{mn}/(2n+a-1)a(a+1)\ldots(a+n)$.

PART II

CHAPTER 5

RELATIONS OF THE BP TO CLASSICAL ORTHONORMAL POLYNOMIALS AND TO OTHER FUNCTIONS.

1. In Chapter 2, the BP have been defined as the polynomial factors of the Bessel functions $K_\nu(z)$ of half-integral order. In fact, it follows from (2.5) with $\nu = n + 1/2$ and $c_0 = (2n)!/2^n n!$ that

$$\theta_n(z) = \theta(z,n+1/2) = c_0 2^{1/2-n}(\Gamma(n+1/2))^{-1} z^{n+1/2} e^z K_{n+1/2}(z),$$

so that

(1) $$\theta_n(z) = \sqrt{\frac{2z}{\pi}}\; z^n e^z K_{n+1/2}(z).$$

Many of the properties of BP can be obtained in the easiest way from (1), by using known properties of the much studied functions $K_{n+1/2}(z)$. This, however, is not the only possible point of view. The BP may also be considered as particular cases of hypergeometric functions, (see, e.g., [90], [2], [3]) of Laguerre polynomials [3], of Whittaker functions [2], or as limiting cases of Jacobi polynomials [2]. They also may be defined as partial quotients of certain continued fractions (see [83]), or, as seen in Chapter 3, as determinants of some simple matrices (see [85]).

While the identity of all these definitions is easy to establish, some of them are more convenient than others for the proof of specific properties of the BP.

In the following sections of the present chapter we shall present the connection of the BP with hypergeometric functions, with Jacobi and Laguerre polynomials, with Lommel Polynomials and with the Whittaker functions. The connections with continued fractions will be discussed in Chapter 8.

These connections with well studied functions have been used by many authors, in order to prove specific properties of the BP in the most convenient way.

2. Let

$${}_pF_q(a_1,a_2,\ldots,a_p;\, b_1,b_2,\ldots,b_q;z) = 1+\sum_{m=1}^{\infty} \frac{(a_1)_m(a_2)_m\cdots(a_p)_m}{(b_1)_m(b_2)_m\cdots(b_q)_m}\;\frac{z^n}{m!},$$

where $(a_j)_m = a_j(a_j+1)(a_j+2)\ldots(a_j+m-1)$, and $b_1b_2\ldots b_q \neq 0$.

The conditions of convergence of this sum are well known (see, e.g. [90], or [6]), but we do not need them here, because we are interested only in the polynomial case of terminating series. The functions ${}_pF_q$ represented by convergent (in particular, by terminating) series are the <u>generalized hypergeometric functions</u>.

It may be verified (see [90]) that ${}_pF_q$ is a solution of the differential equation

(2) $$[\delta \prod_{j=1}^{q} (\delta+b_j-1)-z \prod_{i=1}^{p} (\delta+a_i)]w = 0,$$

where $\delta = z \frac{d}{dz}$ as in Chapter 2.

In general, if $p > q+1$, the corresponding series diverges for all $z \neq 0$; if, however, one of the a_i's is a negative integer, then the series reduces to a polynomial and (2) is meaningful without any restriction on the non-negative integers p and q.

In particular, for $q = 0$ (i.e., if $(b_1)_n(b_2)_n\ldots(b_q)_n = 1$ for all n), $a_1 = -n$, and $a_2 = 1+n$, the general term of ${}_pF_q$ becomes

$$(-n)(1-n)\ldots(m-1-n)\cdot(n+1)(n+2)\ldots(n+m)\,\frac{z^m}{m!} = (-1)^m \frac{(n+m)!}{(n-m)!}\,\frac{z^m}{m!},$$

so that the general term of ${}_2F_0(-n,1+n;-;-z/2)$ reads $\frac{(n+m)!}{2^m(n-m)!m!}\,z^m$.

Comparison with (2.8) shows that

(3) $$y_n(z) = {}_2F_0(-n,1+n;-;-z/2).$$

We also verify that in this case (2) reduces to $[\delta-z(\delta-n)(\delta+n+1)]w = 0$, or upon replacing z by $-z/2$, to $[\delta+(z/2)(\delta-n)(\delta+n+1)]w = 0$, which is (2.25). In the same way one verifies (see [89], [90] and [50]) that

(4) $$y_n(z;a,b) = {}_2F_0(-n,n+a-1;-;-z/b).$$

Just as the well-known theory of Bessel functions can be used to obtain properties of BP from (1) so the equally well-known theory of hypergeometric function can be used for the same purpose, on account of (3).

This has indeed been done successfully (see, e.g. [3], [2], [112], [90], [15], [47], [88], [89], [95]). In this context, it is convenient to recall also that

(5) $$\lim_{X\to\infty} {}_2F_1(a_1,a_2;X;Xz) = {}_2F_0(a_1,a_2;-;z).$$

This follows immediately from the fact that, for fixed n,

$$\lim_{X\to\infty} \frac{(a_1)_n(a_2)_n}{(X)_n}\,\frac{X^n x^n}{n!} = \frac{(a_1)_n(a_2)_n x^n}{n!}.$$

3. We recall that the Jacobi Polynomial $P_n^{(\alpha,\beta)}(z)$ may be defined (see [90], p. 254) by

$$P_n^{(\alpha,\beta)}(z) = \frac{(1+\alpha)_n}{n!}\,{}_2F_1(-n,1+\alpha+\beta+n;1+\alpha;\,(1-z)/2).$$

If we set here $\alpha+1 = X$, $z = 1+2X(y/b)$, $X+\beta = a-1$, then $\alpha = X-1$, $\beta = a-1-X$, $(1-z)/2 = -X(y/b)$ and

$$
{}_2F_1(-n,1+\alpha+\beta+n;1+\alpha;(1-z)/2) = {}_2F_1(-n,a+n-1;X;-X(y/b))
$$

$$
= \frac{n!}{(1+\alpha)_n} P_n^{(X-1,a-X-1)}(1+2X(y/b)) = \frac{\Gamma(\alpha+1)n!}{\Gamma(n+\alpha+1)} P_n^{(X-1,a-X-1)}(1+2X(y/b))
$$

$$
= n \frac{\Gamma(X)\Gamma(n)}{\Gamma(n+X)} P_n^{(X-1,a-X-1)}(1+2X(y/b)).
$$

If we take the limit for $X \to \infty$, then the left hand side becomes (see (5)) ${}_2F_0(-n,a+n-1;-;-(y/b))$, or, by (4), $y_n(y;a,b)$. We have obtained the result (see [2]) (with z instead of y):

$$
(6) \qquad y_n(z;a,b) = \lim_{X \to \infty} nB(n,X)P_n^{(X-1,a-X-1)}(1+2X(z/b)),
$$

where as usually $B(a,b) = \dfrac{\Gamma(a)\Gamma(b)}{\Gamma(a+b)}$.

By using the classical differential equation, Rodrigues formula, recurrence formulae, generating function, etc. of the Jacobi polynomials, one easily obtains the corresponding results for $y_n(z;a,b)$, by passage to the limit in (6). See [2] for details.

4. Laguerre Polynomials may be defined (see [90], p. 201) by

$$
L_n^{(\alpha)}(x) = \sum_{m=0}^{n} \frac{(-1)^m(1+\alpha)_n x^m}{m!(n-m)!(1+\alpha)_m} .
$$

It follows that $L_n^{(-2n-a+1)}(2/z) = \sum_{m=0}^{n} (-1)^m \dfrac{(-2n-a+2)_n 2^m}{m!(n-m)!(-2n+2-a)_m} z^{-m}$

and

$$
n!(-z/2)^n L_n^{(-2n-a+1)}(2/z) =
$$

$$
= \sum_{m=0}^{n} (-1)^{m+n} 2^{m-n} \frac{(-2n-a+2)(-2n-a+3)\dots(-n-a+1)n!z^{n-m}}{m!(n-m)!(-2n+2-a)(-2n+3-a)\dots(-2n+m-a+1)}
$$

$$
= \sum_{m=0}^{n} \frac{(2n+a-2)(2n+a-3)\dots(n+a-1)n!}{m!(n-m)!(2n+a-2)\dots(2n+a-m-1)} (z/2)^{n-m}
$$

$$
= \sum_{m=0}^{n} \frac{(2n+a-m-2)\dots(n+a-1)n!}{m!(n-m)!} (z/2)^{n-m}
$$

$$
= \sum_{k=0}^{n} \frac{(n+k+a-2)\dots(n+a-1)n!}{k!(n-k)!2^k} z^k .
$$

Comparison with (2.27) now shows (see [3]) that

$$y_n(z;a) = n!(-z/2)^n L_n^{(-2n-a+1)}(2/z).$$

Finally, if we replace z/2 by z/b, we obtain

$$y_n(z;a,b) = n!(-z/b)^n L_n^{(-2n-a+1)}(b/z),$$

and

$$\theta_n(z;a,b) = z^n y_n(z^{-1};a,b) = (-1)^n n! b^{-n} L_n^{(-2n-a+1)}(bz).$$

In particular (see [19], [3])

$$\theta_n(z) = (-1)^n n! 2^{-n} L_n^{(-2n-1)}(2z).$$

5. The Whittaker functions $W_{k,m}(z)$ are solutions of the differential equation $w''+\{-1/4+kz^{-1}+(1/4-m^2)z^{-2}\}w = 0$ and may be represented for $z \neq 0$ (an empty product being set equal to one) by (see [67]):

$$W_{k,m}(z) = e^{-z/2} z^k \sum_{r=0}^{q} \frac{(m^2-(k-1/2)^2)(m^2-(k-3/2)^2)\ldots(m^2-(k+1/2-r)^2)}{r^2 z^r} + R;$$

here $R = R(z;m,k,q) \to 0$ for fixed q and $|z| \to \infty$.

The series obtained by letting $q \to \infty$ is asymptotic. However, if $m+k = n+1/2$ with integral n, then the sum is terminating and $R = 0$. Indeed $m^2-(k+1/2-r)^2 = (m+k+1/2-r)(m-k-1/2+r)$ vanishes for $r = n+1$; hence,

$$W_{k,-k+1/2+n}(z)$$

$$= e^{-z/2} z^k \sum_{r=0}^{n} \frac{n(n+1-2k)(n-1)(n+2-2k)\ldots(n-r+1)(n+r-2k)}{r! z^r}.$$

One recognizes in the sum precisely ${}_2F_0(-n,n+1-2k;-;(-z^{-1}))$, or, by (4), $y_n(z^{-1};2(1-k),1)$. It follows that $y_n(z^{-1};2(1-k),1) = e^{z/2} z^{-k} W_{k,-k+1/2+n}(z)$, or writing z^{-1} for z and a for $2(1-k)$, $y_n(z;a,1) = e^{1/2z} z^{1-a/2} W_{1-a/2,(a-1)/2+n}(1/z)$.

Finally, if we replace on the right z by z/b we obtain (see [2] and [5]; observe, however, misprint in (13) of [2]):

$$y_n(z;a,b) = e^{b/2z}(z/b)^{1-a/2} W_{1-a/2,(a-1)/2+n}(b/z).$$

For a = b = 2, one obtains the ordinary BP:

$$y_n(z) = e^{1/z} W_{0,n+1/2}(2/z).$$

6. <u>The Lommel Polynomials</u> $R_{n,\nu}(z)$ may be defined (see [61], §9.6-9.73) by

$$R_{n,\nu}(z) = \frac{1}{2}\pi z \operatorname{cosec} \pi\nu \{J_{n+\nu}(z)J_{-\nu+1}(z)+(-1)^n J_{-\nu-n}(z)J_{\nu-1}(z)\}.$$

If one replaces here the functions $J_m(z)$, for $m = k + \frac{1}{2}$, half an odd integer, by $J_{k+1/2}(z) = (2\pi z)^{-1/2}\{i^{-n-1}e^{iz}y_n(i/z) + i^{n+1}e^{-iz}y_n(-i/z)\}$ (see [68], (11)) one obtains (see Dickinson [47]): For $\nu = s + \frac{1}{2}$, $R_{n,\nu}(z^{-1}) =$ $\frac{1}{2}\{i^{-n}y_{n+s}(iz)y_{s-1}(-iz)+i^n y_{n+s}(-iz)y_{s-1}(iz)\}$. The relation between the Bessel function $J_{k+1/2}(z)$ and the BP $y_n(\pm i/z)$ follows (see Chapters 2 and 3), e.g., by $K_{n+1/2}(z^{-1}) = (\pi z/2)^{1/2}e^{-1/z}y_n(z)$ and the relations between $K_\nu(z)$ and $J_\nu(z)$, or, as in [68], directly from the differential equations.

7. We formalize some of the results obtained so far in this chapter in the following theorem.

<u>THEOREM 1.</u> *With previous notations for hypergeometric functions, Jacobi Polynomials, Beta Function, Laguerre Polynomials and Whittaker functions, the following relations (among others) connect the BP with these special functions:*

(i) $y_n(z;a,b) = {}_2F_0(-n,n+a-1; -; -z/b)$;

(ii) $y_n(z) = {}_2F_0(-n,n+1;-;-z/2)$;

(iii) $y_n(z;a,b) = \lim_{x\to\infty} nB(n,x)P^{(x-1,a-x-1)}(1+2x(y/b))$;

(iv) $y_n(z) = \lim_{x\to\infty} nB(n,x)P_n^{(x-1,1-x)}(1+xy)$;

(v) $y_n(z;a,b) = n!(-z/b)^n L_n^{(-2n-a+1)}(b/z)$;

(vi) $y_n(z) = n!(-z/2)^n L_n^{(-2n-1)}(2/z)$;

(vii) $y_n(z;a,b) = e^{b/2z}(z/b)^{1-a/2}W_{1-a/2,\, \frac{1}{2}(a-1)+n}(b/z)$;

(viii) $y_n(z) = e^{1/z}\, W_{0,n+\frac{1}{2}}(2/z)$;

(ix) $\theta_n(z;a,b) = (-1)^n n! b^{-n} L_n^{(-2n-a+1)}(bz)$;

(x) $\theta_n(z) = (-1)^n n! 2^{-n} L_n^{(-2n-1)}(2z)$.

<u>REMARK.</u> In the statement of the theorem all relations referring to the case $a \neq b = 2$ have been omitted, because these are immediately obtainable by setting $b = 2$ in the formulae for $y_n(z;a,b)$ and $\theta_n(z;a,b)$. The formulae for $y_n(z)$ and $\theta_n(z)$ are indicated, because of their rather frequent use.

8. In the preceding sections we have discussed the relations of BP to other special functions essentially from the point of view of identifying BP with these other functions, for particular values of their parameters.

A different kind of connection is pointed out by the following remark that, apparently, was never made before.

Let $P_n(x)$ be the n-th Legendre Polynomial. This is, of course, a particular case of Jacobi Polynomials, but can also be defined by the generating function

$$(1-2zx+z^2)^{-1/2} = \sum_{n=0}^{\infty} P_n(x)z^n. \tag{7}$$

If $P_n^{(r)}(x)$ stands for the r-th derivative of $P_n(x)$, then the following theorem holds

<u>THEOREM 2.</u> $y_n(z) = \sum_{r=0}^{n} P_n^{(r)}(1)z^r;$

$$\theta_n(z) = \sum_{r=0}^{n} P_n^{(r)}(1)z^{n-r}.$$

The two statements are, of course, equivalent, on account of (2.10). Both are immediate consequences of

<u>LEMMA 1.</u> $a_r^{(n)} = P_n^{(r)}(1).$

On account of (2.8), Lemma 1 is equivalent to

<u>LEMMA 2.</u> $P_n^{(r)}(1) = \dfrac{(n+r)!}{2^r r!(n-r)!}.$

This lemma appears to have been known to I. Schur, but this writer is unable to locate a reference in Schur's work. However, many essentially distinct proofs are known, by N. du Plessis, [48], E. Rainville [50], I.N. Herstein [31], and this writer [23], among others. Here is the particular simple one due to du Plessis: We differentiate (7) r times with respect to x and obtain

$$z^r.1.3....(2r-1)(1-2zx+z^2)^{-r-1/2} = \sum_{n=0}^{\infty} P_n^{(r)}(x)z^n. \tag{8}$$

By setting x = 1 and expanding the binomial, we obtain on the left of (8)

$$z^r.1.3...(2r-1)(1-z)^{-2r-1} = z^r.1.3...(2r-1)\sum_{m=0}^{\infty}\binom{-2r-1}{m}(-z)^m$$

$$= 1.3...(2r-1)\sum_{m=0}^{\infty}(-1)^m\binom{-2r-1}{m} z^{m+r} =$$

$$1.3...(2r-1)\sum_{m=0}^{\infty}\frac{(2r+1)(2r+2)...(2r+m)}{m!} z^{m+r} =$$

$$1.3\ldots(2r-1)\sum_{n=r}^{\infty}\frac{(n+r)(n+r-1)\ldots(2r+1)}{(n-r)!}\;z^n.$$

For $x = 1$, the right hand side of (8) becomes $\sum_{n=0}^{\infty} P_n^{(r)}(1)z^n$ and Lemma 2, hence Lemma 1 and Theorem 2 follow by comparison of the coefficients of z^n in (8).

CHAPTER 6

GENERATING FUNCTIONS

1. A very large number of generating functions for BP are known. Some are formal series, others represent functions within certain domains of convergence; some are obtained by more or less classical methods, others are obtained by ad-hoc considerations, while some others still are obtained by the theory of Lie groups. An attempt is made to list here at least the more important, or more useful of the known generating function, but even for many of the listed ones it will be necessary to refer the reader to the original literature for complete proofs.

2. Let us consider the function $f(t,z) = \exp\{[1-(1-2zt)^{1/2}]/z\}$. By logarithmic differentiation with respect to t we obtain

$\frac{f'}{f}(t,z) = (1-2zt)^{-1/2}$, so that $f'(t,z) = (1-2zt)^{-1/2} f(t,z)$, and by a further differentiation $f''(t,z) = \{(1-2zt)^{-1}+z(1-2zt)^{-3/2}\} f(t,z)$

$$= \{a_0^{(1)}(1-2zt)^{-1}+a_1^{(1)} z(1-2zt)^{-3/2}\} f(t,z),$$

with $a_0^{(1)} = a_1^{(1)} = 1$, the two coefficients of $y_1(z)$.

These formulae are the instances $k = 1$ and $k = 2$ of the

PROPOSITION. *For* $k \geq 1$,

$$(1) \qquad f^{(k)}(t,z) = f(z,t) \sum_{m=0}^{k-1} a_m^{(k-1)} z^m (1-2zt)^{-(k+m)/2},$$

where $f^{(k)}(t,z) = \frac{\partial^k f}{\partial t^k}$ *and where the* $a_m^{(j)}$ *are the coefficients of* $y_j(z)$, *as given by* (2.8).

Proof. We already know that (1) holds for $k = 1$ and $k = 2$, and proceed to complete the proof by induction on k.

We differentiate (1) and obtain:

$$f^{(k+1)}(t,z) = \sum_{m=0}^{k-1} a_m^{(k-1)} z^m (1-2zt)^{-(k+m)/2} \cdot (1-2zt)^{-1/2} f(t,z)$$

$$+ f(t,z) \sum_{m=0}^{k-1} a_m^{(k-1)} z^{m+1} (k+m)(1-2zt)^{-(k+m+2)/2}$$

$$= f(t,z) \sum_{m=0}^{k} (a_m^{(k-1)} + (k+m-1) a_{m-1}^{(k-1)} z^m (1-2zt)^{-(k+m+1)/2}.$$

By using (3.19), this can be written as

$$f^{(k+1)}(t,z) = f(t,z) \sum_{m=0}^{k} a_m^{(k)} z^m (1-2zt)^{-(k+m+1)/2}$$

and this is precisely (1), with k+1 instead of k. The Proposition is proved.

As immediate consequences we note the following corollaries.

<u>COROLLARY 1</u>. $f^{(k+1)}(0,z) = y_k(z)$.

<u>COROLLARY 2</u>. (see [68]). $f(t,z) = \sum_{k=0}^{\infty} t^k y_{k-1}(z)/k!$,

where (see (2.12)), *for* $k = 0$, $y_{-1}(z) = y_0(z) = 1$.

It follows that the function $f(t,z)$ is a generating function for the polynomials $y_{k-1}(z)$. By differentiation with respect to t we obtain (see [17]):

$$\frac{\partial f}{\partial t} = (1-2tz)^{-1/2} f(t,z) = \sum_{k=0}^{\infty} t^k y_k(z)/k! \ . \tag{2}$$

The function $\partial f/\partial t$ is, therefore, a generating function for the $y_k(z)$.

3. If we replace in (2), $y_k(z)$ by $z^k \theta_k(z^{-1})$, we obtain

$$(1-2tz)^{-1/2} \exp\{[1-(1-2tz)^{1/2}]/z\} = \sum_{k=0}^{\infty} t^k z^k \theta_k(z^{-1})/k! \ ,$$

or, replacing z by z^{-1}:

$$(1-2tz^{-1})^{-1/2} \exp\{z(1-(1-2tz^{-1})^{1/2}\} = \sum_{k=0}^{\infty} (tz^{-1})^k \theta_k(z)/k! \ .$$

We now denote tz^{-1} by v and this leads to

$$(1-2v)^{-1/2} \exp\{z(1-(1-2v)^{1/2})\} = \sum_{k=0}^{\infty} v^k \theta_k(z)/k! \ . \tag{3}$$

We may also, following Burchnall [17], set $1-(1-2v)^{1/2} = 2u$. If we substitute this in (3) we obtain (see [17])

$$(1-2u)^{-1} e^{2zu} = \sum_{k=0}^{\infty} \{2u(1-u)\}^k \theta_k(z)/k! \ . \tag{4}$$

Burchnall calls this a pseudo-generating function for $\theta_k(z)$.

The corresponding pseudo-generating function for the generalized BP $\theta_k(z;a,b)$ is (see [17] and also [2]).

$$(1-2u)^{-1}(1-u)^{2-a} e^{bzu} = \sum_{k=0}^{\infty} \{bu(1-u)\}^k \theta_k(z;a,b)/k! \ . \tag{5}$$

By replacing here $\theta_k(z;a,b)$ by $z^k y_k(z^{-1};a,b)$ and then z by z^{-1}, we obtain

$$(1-2u)^{-1}(1-u)^{2-a} e^{bu/z} = \sum_{k=0}^{\infty} \{buz^{-1}(1-u)\}^k y_k(z;a,b)/k!,$$ or, with $uz^{-1} = v$,

$$(1-2vz)^{-1}(1-vz)^{2-a}e^{bv} = \sum_{k=0}^{\infty} \{bv(1-vz)\}^k y_k(z;a,b)/k! .$$

Finally, still following Burchnall [17], let $2v(1-vz) = t$, so that $1-2tz = (1-2vz)^2$ and we obtain the following generating function for the polynomials $y_k(z;a,b)$:

(6) $$(1-2tz)^{-1/2}\{[1+(1-2zt)^{1/2}]/2\}^{2-a}\exp\{(b/2z)(1-(1-2tz)^{1/2})\}$$

$$= \sum_{k=0}^{\infty} (b/2)^k t^k y_k(z;a,b)/k! .$$

A simple proof of (6) is also given in [43] and [2] and, for $b = 2$ in [16], see also [50].

4. Burchnall's generating function has been generalized by Rainville (see [89] and [90]), as follows:

Let

$$\psi_n(x) = {}_{q+2}F_p(-n,c+n,1-\beta_1-n,\ldots,1-\beta_q-n;\ 1-\alpha_1-n,1-\alpha_2-n,\ldots,1-\alpha_p-n;(-1)^{p+q+1}x);$$

then $\psi_n(x)$ admits a generating function of the form

$$(1-4xt)^{-1/2}\{\tfrac{1}{2}\ [1+(1-4xt)^{1/2}]\}^{1-c}\ {}_pF_q(\alpha_1,\alpha_2,\ldots,\alpha_p;\beta_1,\beta_2,\ldots,\beta_q;[2t(1+(1-4xt)^{1/2})^{-1}])$$

(7) $$= \sum_{n=0}^{\infty} \frac{\psi_n(z)t^n}{n!}\ \frac{(\alpha_1)_n\cdots(\alpha_p)_n}{(\beta_1)_n\cdots(\beta_q)_n} .$$

Burchnall's generating function (6) is essentially (7) with $p = q = 0$, $c = a-1$, $x = zb^{-1}$, and t replaced by $bt/2$.

A different type of generating functions for BP is obtained by Rainville [89] as follows: Set $\sigma_n(x) = {}_{p+2}F_q(-n,c+n,\alpha_1,\ldots,\alpha_p;\ \beta_1,\ldots,\beta_q;-x)$; then

(8) $$(1-t)^{-c}\ {}_{p+2}F_q(c/2,(c+1)/2,\alpha_1,\ldots,\alpha_p;\beta_1,\ldots,\beta_q;\ 4xt(1-t)^{-2})$$

$$= \sum_{n=0}^{\infty} \sigma_n(x)(c)_n t^n/n! .$$

For $p = q = 0$, $\sigma_n(x) = {}_2F_0(-n,c+n;-;-x) = y_n(bx,c+1,b)$, so that (8) is the generating function for the $y_n(z;a,b)$, with $z = bx$, $a = c+1$, and the same b.

5. W.A. Al-Salam [3] uses two generating functions for Jacobi Polynomials due to Feldheim [**16**] and the relation between BP and Jacobi Polynomials (see Chapter 5),

in order to obtain (in a slightly different notation and somewhat lesser generality) the generating function

$$(9)\qquad e^t(1-tz/b)^{1-\alpha} = \sum_{n=0}^{\infty} t^n y_n(z;\alpha-n,b)/n!\ ,$$

and the formal (divergent) generating function

$$(10)\qquad (1-t)^{-1}{}_2F_0(a-2,1;-;tz/b(1-t)) \sim \sum_{n=0}^{\infty} t^n y_n(z;a-n-1,b).$$

On account of (5.4) for integral a (10) may be written also as

$$(10')\qquad (1-t)^{-1}y_{2-a}(z/(1-t^{-1});a,b) \sim \sum_{n=0}^{\infty} t^n y_n(z;a-n-1,b).$$

By using the connection with Laguerre Polynomials (see Chapter 5) and known generating functions of these polynomials, Al-Salam [3] also obtains

$$(11)\qquad (1+tz/b)^{\alpha-2}\exp\{t/(1+tz/b)\} = \sum_{n=0}^{\infty} t^n y_n(z;\alpha-2n,b)/n!\ .$$

The characteristic feature of these generating functions is that they lead to sums of BP $y_n(z;a,b)$, where the entry $a = a(n)$, and is not a constant.

Next, the relation $L_n^{(\alpha)}(x+t) = e^t \sum_{k=0}^{\infty} (-1)^k t^k L_n^{(\alpha+k)}(x)/k!$ leads (see (6.10) in [3]) to

$$(12)\qquad e^t(1-tz/b)^n y_n(Z;a,c) = \sum_{k=0}^{\infty} t^k y_n(z;a-k,b)/k!\ ,$$

where $Z = cz/(b-tz)$.

One may observe that the parameter c in the first member is idle and can be replaced by 2, provided that one then sets $Z = 2z/(b-tz)$. It also may be worthwhile to observe, that in the second member the subscript of all BP is n, while the "a"-parameter varies, depending on k and the sum is over k.

Al-Salam gives also several, rather complicated bilinear generating functions, as well as the formal (no convergence) generating function

$$(13)\qquad \sum_{n=0}^{\infty} \binom{n+a-2}{n} t^n y_n(z;a,b) \sim (1-t)^{1-a}{}_2F_0\left(\frac{a-1}{2}, \frac{a}{2}; -; \frac{bzt}{(1-t)^2}\right).$$

In case $(1-a)/2 = m$ is an integer, this is equivalent (by (5.4)) to

$$(13')\qquad \sum_{n=0}^{\infty} \binom{n-2m-1}{n} t^n y_n(z;1-2m,b) \sim (1-t)^{2m} y_m(-4zt(1-t)^{-2}; \tfrac{3}{2}-2m,b).$$

For $a = 2$, on the other hand, (13) reduces (see [3]) to

$$(13'')\qquad \sum_{n=0}^{\infty} y_n(z;2,b)t^n \sim (1-t)^{-1}{}_2F_0\left(\tfrac{1}{2};1;-;\frac{bzt}{(1-t)^2}\right),$$

essentially obtained already by Krall and Frink [68].

From these formulae corresponding ones for $\theta_n(z;a,b)$ can be obtained by simple manipulations.

A somewhat different generating function, involving Hermite polynomials is found in [3], (6.13). It can be obtained by setting a = 1-k in (13) and interchanging n and k. With some simplifications (also some misprints in [3] need correction), this reads

$$\sum_{k=0}^{\infty} (-1)^k \binom{n}{k} t^k y_k(z;1-k,b) = e^{-\pi i n/2}(tz/b)^{n/2} H_n\left(\frac{i(1-t)}{2}\left(\frac{b}{zt}\right)^{1/2}\right).$$

6. Brafman [15] considers the Cauchy product of the series representations for ${}_2F_0(\alpha,c-\alpha;-;\frac{1}{2}(t+(t^2-4xt)^{1/2})$ and ${}_2F_0(\alpha,c-\alpha;-;\frac{1}{2}(t-(t^2-4xt)^{1/2})$, obtains formally a sum over ${}_3F_2$'s, transforms the latter by use of a formula of Whipple [65] and obtains in this way the formal result

$${}_2F_0(\alpha,c-\alpha;-;\tfrac{1}{2}(t-(t^2-4xt)^{1/2}))\,{}_2F_0(\alpha,c-\alpha;-;\tfrac{1}{2}(t+(t^2-4xt)^{1/2})) \sim \sum_{n=0}^{\infty} ((\alpha)_n(c-\alpha)_n/n!)t^n\,{}_2F_0(-n,c+n;-;x).$$

By observing that ${}_2F_0(-n,c+n;-;x) = y_n(-bx;c+1,b)$ we obtain the formal generating functions for BP:

(14) $${}_2F_0(\alpha,c-\alpha;-;\tfrac{1}{2}(t-(t^2-4xt)^{1/2}))\,{}_2F_0(\alpha,c-\alpha;-;\tfrac{1}{2}(t+(t^2-4xt)^{1/2})) \sim \sum_{n=0}^{\infty} c_n t^n y_n(-bx,c+1,b),$$

where $c_n = c_n(\alpha,c) = (\alpha)_n(c-\alpha)_n/n!$. If $\alpha = -m$, a negative integer, then also the first member of (14) can be written as a product of BP, namely

$$y_m(z_1,c+1,2)y_m(z_2,c+1,2),$$

where $z_1 = -t+(t^2-4xt)^{1/2}$, $z_2 = -t-(t^2-4xt)^{1/2}$.

We may observe that in this case the second member reduces to a finite sum, because now $(\alpha)_n = (-m)(1-m)\dots(n-m-1)$ and vanishes for $n > m$, and the formal equality becomes an actual one (see [3] in somewhat different notations) and reads:

(15) $$y_m(z_1,a,2)y_m(z_2,a,2) = \sum_{n=0}^{m} \binom{m-n+1}{n}(a+m-n)_n t^n y_n(-bx,a,b),$$

with previous values for z_1 and z_2.

7. Carlitz [19] starts from Corollary 2, sets $f_n(z) = z^n y_{n-1}(z^{-1}) = z\theta_{n-1}(z)$ and observes that Corollary 2 can be written as

$$\exp\{z(1-(1-2t)^{1/2})\} = \sum_{n=0}^{\infty} f_n(z)t^n/n! \ , \tag{16}$$

or, if we set $1-2t = (1-2u)^2$,

$$e^{2uz} = \sum_{n=0}^{\infty} 2^n f_n(z)(u-u^2)^n/n! \ . \tag{17}$$

If we replace in (17), z by z^{-1}, so that $f_n(z^{-1}) = z^n y_{n-1}(z)$ and then set $2uz^{-1} = x$, (17) becomes (see [68])

$$e^z = \sum_{n=0}^{\infty} (z-xz^2/2)^n y_{n-1}(x)/n! \ . \tag{18}$$

In this surprizing formula, the right hand side is in fact independent of x and, for $x = 0$, it reduces trivially to the left hand side.

By using Brafman's formal equality (14) in the case $c = 1$, Dickinson [47] obtains a generating function for the modified Lommel polynomials.

Here we have reached a point, where it is debatable, whether we are still dealing with generating functions in their original sense. Formulae like (17), or (18), may be considered more properly as belonging to the theory of representation of functions by series of BP, a topic that will be discussed in Chapter 9.

8. This chapter would be incomplete without the mention of a most important general method that permits us to obtain a large number of generating functions, among them most of the previous ones. It is based on the theory of Lie groups. It is, unfortunately, not possible to develop here the complete theory and the interested reader is advised to consult, e.g., the excellent presentation of the method in [73]. The method has been introduced by L. Weisner [62], [63], [64] and probably the clearest presentation of the basic idea is still the one of [62]. On the other hand, the only complete presentation known to the present author of all details needed for the non-specialist is to be found in E.B. McBride [73]. A somewhat different approach can be found in the work of S.K. Chatterjea [35], and M.K. Das [43].

Following Weisner [62], we consider a linear, ordinary differential equation, that depends also linearly on a (not necessarily integral) parameter α, say

$$L(x, \frac{d}{dx}, \alpha)v = 0, \tag{19}$$

with $\alpha \in A$, say, where A is some denumerable set.

If y is another independent variable, then for every β, with $v = v_\alpha(x)$, also $u(x,y) = y^\beta v_\alpha(x)$ is a solution of (19). Let now, in general, $u = u(x,y)$ be a function of the two independent variables x,y that satisfies $\delta_y u = \alpha u$; here we have set $\delta_y = y\frac{\partial}{\partial y}$, in analogy with the notation of Chapter 2. By the linearity of $L(x, \frac{d}{dx}, \alpha)$ in α, and because $\delta_y y^\alpha = \alpha y^\alpha$, it follows that a function u(x,y), that satisfies $\delta_y u = \alpha u$, and also

(20) $$Lu = L(x, \frac{\partial}{\partial x}, \delta_y)u = 0,$$

is of the form $u(x,y) = y^\alpha v_\alpha(x)$, with $v_\alpha(x)$ a solution of (19). The converse is fairly obvious: if $v_\alpha(x)$ is a solution of (19), then $u(x,y) = y^\alpha v_\alpha(x)$ is a solution of the system $\delta_y u = \alpha u$, $Lu = 0$. Hence, if $\{v_\alpha(x)\}$ is a family of solutions of (19), then the se. $\{y^\alpha v_\alpha(x)\}$ is a family of solutions of (20).

Let us now assume that independently of what precedes we can

(i) find a solution $u = g(x,y)$ of (20); and

(ii) expand $g(x,y)$ in a series of the form

(21) $$g(x,y) = \sum_{\alpha \varepsilon A} g_\alpha(x) y^\alpha.$$

If, furthermore, the series converges sufficiently well so that a termwise application of the operator $L = L(x, \frac{\partial}{\partial x}, \delta_y)$ can be justified, it follows that

$L(x, \frac{\partial}{\partial x}, \delta_y)(g_\alpha(x)y^\alpha) = 0$, whence $L(x, \frac{d}{dx}, \alpha)(g_\alpha(x)y^\alpha) = y^\alpha L(x, \frac{d}{dx}, \alpha) g_\alpha(x) = 0$,

so that $\{g_\alpha(x)\}$ is a family of solutions of (19). In this way we have obtained, according to (21), a generating function $g(x,y)$ for a family $\{g_\alpha(x)\}_{\alpha \varepsilon A}$ of solutions of (19).

The difficulty of this approach consists in the fact that most usual methods for the solution of (20) lead to solutions of the form $u = g(x,y) = g_\alpha(x)y^\alpha$ and sums of such solutions. This, of course, reduces (21) to a trivial identity. If, however, solutions $u = g(x,y)$ of (20) can be found in closed form, not of the type $\sum_\alpha g_\alpha(x)y^\alpha$, then (21) yields genuine generating functions for the class $\{g_\alpha(x)\}_{\alpha \varepsilon A}$ of solutions of (19). Moreover, distinct (and linearly independent) solutions $g(x,y)$ of (20) furnish distinct generating functions.

A particular case in which non-trivial solutions of (20) can be found is that when (20) admits a non-trivial group of transformations (it always admits the groups $x \to \xi$, $y \to t\eta$). Sometimes it is possible to guess a set of differential operators of the first order and verify on hand of their commutation relations that they generate a Lie group and that these generators commute with the operator L, or at least with $\psi(x) \cdot L$, for an appropriate function $\psi(x)$. In the particular case of the classical orthogonal polynomials, or of Bessel functions, one may take as parameter α the degree, or order n and attempt to use as generators of the Lie group the operators that raise, or lower n. A variety of special methods, shortcuts, etc. have been used by various authors, in order to obtain their results (see, e.g. [35], [43]).

In general, however, a rather careful study of the Lie algebra, its commutation relations, conjugacy classes of operators, etc., is needed (see, e.g. [62], [63], [64], [73]).

As an example (see [73] and [43]), let $u_n(x) = x^{1/2}e^{-1/x}y_n(x)$, where $y_n(x)$ is the n-th BP, and set $F_n(x,t) = e^{(n+1/2)t}u_n(x)$. (We write t rather than y, to avoid confusion with $y_n(x)$). We observe that $\underline{R} = e^t(x \frac{\partial}{\partial y} + x^2 \frac{\partial}{\partial x})$ is a right shift operator for F_n. Indeed,

$$\underline{R}F_n = e^t\{x(n+1/2)e^{(n+1/2)t}u_n + x^2e^{(n+1/2)t}u_n'\}$$

$$= e^{(n+3/2)t}\{(n+1/2)x\cdot x^{1/2}e^{-1/x}y_n + x^2(\tfrac{1}{2}x^{-1/2}e^{-1/x} + x^{-3/2}e^{-1/x})y_n + x^{5/2}e^{-1/x}y_n'\}$$

$$= e^{(n+3/2)t}e^{-1/x}x^{1/2}\{((n+1)x+1)y_n + x^2y_n'\} = e^{(n+1+1/2)t}x^{1/2}e^{-1/x}y_{n+1} = F_{n+1},$$

by (3.16). It follows that

$$(22)\quad e^{\omega\underline{R}}F_m(x,t) = \sum_{n=0}^{\infty} \frac{\omega^n}{n!}\underline{R}^nF_m = \sum_{n=0}^{\infty}\frac{\omega^n}{n!}F_{m+n} = x^{1/2}e^{-1/x}\sum_{n=0}^{\infty}\frac{\omega^n}{n!}e^{(m+n+1/2)t}y_{n+m}(x).$$

On the other hand, $e^{\omega\underline{R}}F_m(x,t) = F_m(e^{\omega\underline{R}}x,\ e^{\omega\underline{R}}t)$.

Here $\underline{R}x = e^t(x\frac{\partial}{\partial t} + x^2\frac{\partial}{\partial x})x = e^tx^2$, $\underline{R}^2x = e^t\{x\frac{\partial}{\partial t} + x^2\frac{\partial}{\partial x}\}(e^tx^2) = 3e^{2t}x^3$,

and, by an easy induction on n,

$$\underline{R}^nx = 1.3.5\ldots(2n-1)e^{nt}x^{n+1}.$$

Consequently, $e^{\omega\underline{R}}x = \sum_{n=0}^{\infty}\frac{1.3\ldots(2n-1)}{n!}\omega^ne^{nt}x^{n+1} = x(1-2x\omega e^t)^{-1/2} = x_1$, say.

Similarly, $e^{\omega\underline{R}}t = t + \sum_{n=1}^{\infty}\frac{\omega^n}{n!}\underline{R}^nt = t + \sum_{n=1}^{\infty}\frac{2.4\ldots(2n-2)}{n!}\omega^nx^ne^{nt}$

$$= t + \frac{1}{2}\sum_{n=1}^{\infty}\frac{(2\omega xe^t)^n}{n} = t - \frac{1}{2}\log(1-2\omega xe^t) = t_1, \text{ say.}$$

Consequently, $F_m(e^{\omega\underline{R}}x,\ e^{\omega\underline{R}}t) = F_m(x_1,t_1) =$

$$\exp\{(m+1/2)(t-1/2\log(1-2\omega xe^t))\}\{x(1-2\omega xe^t)^{-1/2}\}^{1/2}\cdot$$

$$\exp\{-(1-2x\omega e^t)^{1/2}x^{-1}\}y_m(x(1-2x\omega e^t)^{-1/2})$$

$$= \exp\{(m+1/2)t\}(1-2\omega xe^t)^{-(m+1/2)/2}x^{1/2}(1-2x\omega e^t)^{-1/4}$$

$$\cdot\exp\{-(1-2x\omega e^t)^{1/2}x^{-1}\}y_m(x(1-2x\omega e^t)^{-1/2})$$

and (22) becomes, after some obvious simplifications,

$$(23)\qquad (1-2x\omega e^t)^{-(m+1)/2}\exp\{[1-(1-2x\omega e^t)^{1/2}]x^{-1}\}y_m(x(1-2x\omega e^t)^{-1/2}) = \sum_{n=0}^{\infty} \frac{\omega^n e^{nt}}{n!} y_{n+m}(x).$$

In particular, if we set t = 0, we obtain

$$(1-2x\omega)^{-(m+1)/2}\exp\{[1-(1-2x\omega)^{1/2}]x^{-1}\}y_m(x(1-2x\omega)^{-1/2}) = \sum_{n=0}^{\infty} \frac{\omega^n}{n!} y_{n+m}(x).$$

Finally, if we take m = 0 and write t for ω, this becomes

$$(1-2xt)^{-1/2}\exp\{[1-(1-2xt)^{1/2}]x^{-1}\} = \sum_{n=0}^{\infty} \frac{t^n}{n!} y_n(x),$$

which is, of course, precisely the generating function (2) of Burchnall.

Instead of the right shift operator, we could have used, of course, the left shift operator

$$\underline{L} = e^{-t}(-x\frac{\partial}{\partial t} + x^2\frac{\partial}{\partial x}),$$ but we obtain exactly the same result.

On the other hand, we may verify that $[\underline{R},\underline{L}] = \underline{RL} - \underline{LR} = 0$, so that these operators commute and $e^{\underline{K}} = e^{u\underline{R}+v\underline{L}} = e^{u\underline{R}}e^{v\underline{L}}$, with u,v any two independent variables. By proceeding with $\underline{K}$ as we did before with $\underline{R}$, we now obtain (see [43])

$$(1-2xu)^{-(m+1)/2}(1-xv)^{m/2}\exp\{[1-(1-xu)^{1/2}(1-xv)^{1/2}]x^{-1}\}y_m(x(1-2xu)^{-1/2}(1-2xv)^{-1/2}) = \sum_{n,k=0}^{\infty} \frac{u^n v^k}{n!k!} y_{m+n-k}(x).$$

In the second member, for $k > m+n$, $y_{m+n-k}(x)$ has to be interpreted (see (2.12)) as $y_{k-m-n-1}(x)$. The formula simplifies somewhat for m = 0.

The generating function (2) had been obtained also by S.K. Chatterjea [35] by a very similar approach but after he first reduces equation (2.11) to its Sturm-Liouville form

$$\frac{d}{dx}\{p(x)u'\} + \lambda\phi(x)u = 0.$$

In a somewhat different notation and by following very closely the exposition in [73], Ming-Po Chen and Chia-Chin Feng [39] obtain the following generating functions:

$$(1-xt)^n e^{bt} y_n(z;a,b) = \sum_{k=1}^{\infty} \frac{(bt)^k}{k!} y_n(x;a-k,b),$$

with $z = \frac{x}{1-xt}$, which is essentially the same as (12); and

$$y_n(z;a,b) = (1-t)^{n+a-1} \sum_{k=0}^{\infty} \binom{n+k+a-2}{k} t^k y_n(x;a+k,b), \text{ with } z = \frac{x}{1-t}.$$

For a = b = 2, in particular, the latter becomes:

$$y_n(x/(1-t)) = (1-t)^{n-1} \sum_{k=0}^{\infty} \binom{n+k}{k} t^k y_n(x;2+k;2).$$

Finally, with $z - xt(t-\omega x)^{-1}(1-t)^{-1}$,

$$(t-\omega x)^n e^{b\omega/t} y_n(z;a,b) = (1-t)^{n+a-1} \sum_{m=0}^{\infty} \sum_{k=0}^{\infty} \binom{n+k-m+a-2}{k} \frac{(\omega b)^m}{m!} t^{n+k-m} y_n(x,a-m+k,b).$$

9. In addition to the generating functions presented here, one finds either different ones, or different proofs for those discussed here, in the following papers: [3], [109], [46], [36], [76], [77], [78], [79], [16], [30], [31], [44]. See also Chapter 15.

CHAPTER 7

FORMULAS OF RODRIGUES' TYPE

1. In Chapter 2, by use of the differential operator $\delta = z\frac{d}{dz}$, that is by following essentially Burchnall's [17] method, we obtained the Rodrigues' formula (2.40), i.e.:

(1) $$\theta_n(z;a,b) = (-1)^n b^{-n} e^{bz} z^{2n+a-1} \frac{d^n}{dz^n}(z^{-n-a+1}e^{-bz}).$$

We mentioned also the important particular cases of (1), corresponding to $b = 2$, and to $a = b = 2$, which are

(1') $$\theta_n(z;a) = (-1)^n 2^{-n} e^{2z} z^{2n+a-1} \frac{d^n}{dz^n}(z^{-n-a+1}e^{-2z})$$

and

(1'') $$\theta_n(z) = (-1)^n 2^{-n} e^{2z} z^{2n+1} \frac{d^n}{dz^n}(z^{-n-1}e^{-2z}),$$

respectively.

It would appear reasonable to obtain corresponding formulae for $y_n(z;a,b)$ and its particular cases from (1), (1'), (1''), simply by writing $1/z$ for z and then multiplying by z^n. This is, in fact, possible, but the computation of $\{\frac{d^n}{du^n}(u^{-n-a+1}e^{-bu})\}_{u=1/z}$ to which one is led is not entirely trivial (see Section 5). For that reason and also in order to present a variety of approaches we shall start out differently and follow essentially [68].

2. We shall need some known facts, which we state as Propositions and, in order to make the presentation selfcontained, we indicate short proofs of them.

Let C be the unit circle $|z| = 1$ and let $w(z)$ be a weight function. With respect to $w(z)$ we define the k-th moment of the power z^j by

$m_{jk} = \frac{1}{2\pi i}\int_C z^k \cdot z^j \cdot w(z)dz$. With these moments we form the determinants

$$\Delta_n(w) = ||m_{jk}||_0^{n-1} = \begin{vmatrix} m_{00} & m_{01} & \cdots & m_{0,n-1} \\ m_{10} & m_{11} & \cdots & m_{1,n-1} \\ - & - & - & - \\ m_{n-1,0} & m_{n-1,1} & \cdots & m_{n-1,n-1} \end{vmatrix}$$

PROPOSITION 1. *If* $w(z)$ *is such that* $\Delta_n(w) \neq 0$ *for all* n, *then the conditions* $\frac{1}{2\pi i}\int_C w(z)z^k u_n(z)dz = 0$ *for* $k = 0,1,\ldots,n-1$ *define a polynomial* $u_n(z)$ *uniquely, up to an arbitrary multiplicative constant.*

Proof. Let $u_n(z) = \sum_{j=0}^{n} a_j z^j$ with $a_n = 1$; then the integral equals

$\sum_{j=0}^{n} a_j \frac{1}{2\pi i} \int_C z^{k+j} w(z)dz = \sum_{j=0}^{n} a_j m_{jk}$ and the conditions become

$\sum_{j=0}^{n-1} a_j m_{jk} = -a_n m_{nk} = -m_{nk}$ $(k = 0,1,..,n-1)$. The determinant of the coefficients a_j $(j = 0,1,\ldots,n-1)$ is $\Delta_n(w)$ and, by assumption, $\Delta_n(w) \neq 0$. It follows that the a_j's are uniquely determined. If the coefficient a_n is allowed to be arbitrary, then all the a_j's have their previous values multiplied by a_n and Proposition 1 is proved.

PROPOSITION 2. *If* $w(z) = e^{-2/z}$, *then* $m_{jk} = (-1)^{j+k+1} \frac{2^{j+k+1}}{(j+k+1)!}$.

Proof. $m_{jk} = \frac{1}{2\pi i} \int_C z^{k+j} e^{-2/z} dz = \frac{1}{2\pi i} \int_C z^{k+j} \left(\sum_{r=0}^{\infty} \frac{1}{r!} \left(\frac{-2}{z}\right)^r\right)dz$

$$= \sum_{r=0}^{\infty} \frac{(-2)^r}{r!} \frac{1}{2\pi i} \int_C z^{k+j-r} dz.$$

The inversion of the summation and integration is easily justified. All integrals vanish, except one, where $k+j-r = -1$, that is, $r = k+j+1$, and this equals $2\pi i$; Proposition 2 is proved.

PROPOSITION 3. *If* $w(z) = e^{-2/z}$, *then* $\Delta_n(w) \neq 0$ *for all* n.

Proof. It is easy, although somewhat computational to determine the actual value

$$\Delta_n(w) = (-1)^{(3n-1)n/2} 2^{n^2} \prod_{\nu=1}^{n-1} (\nu!) / \prod_{\nu=n}^{2n-1} (\nu!),$$

from which $\Delta_n(w) \neq 0$ follows trivially. For a proof of this formula, see **[25]**.

COROLLARY 1. *The conditions* $\int_C e^{-2/z} z^k u_n(z)dz = 0$ *for* $k = 0,1,\ldots,n-1$ *define a polynomial* $u_n(z)$ *of degree* n *uniquely, up to a multiplicative constant.*

Proof is immediate on account of Propositions 1 and 3.

PROPOSITION 4. $u_n(z) = e^{2/z} \frac{d^n}{dz^n} (z^{2n} e^{-2/z})$ *is a polynomial of exact degree* n, *with constant term* 2^n.

Proof. $\frac{d}{dz} (z^{2n} e^{-2/z}) = 2n\, z^{2n-1} e^{-2/z} + 2z^{2n-2} e^{-2/z} = e^{-2/z} (2n\, z^{2n-1} + 2z^{2n-2})$;

this is the instance m = 1 of the general statement $\frac{d^m}{dz^m}(z^{2n}e^{-2/z}) = e^{-2/z}P_{2n-m}(z)$, where $P_{2n-m}(z)$ is a polynomial of degree 2n-m. One completes the proof by induction on m. Next, by considering Leibniz's Rule for differentiation, we observe that the lowest power of z is obtained from the term $z^{2n}\frac{d^m}{dz^m}e^{-2/z} = z^{2n}\{(2/z^2)^m+\ldots\}e^{-2/z} = (2^m z^{2(n-m)}+\ldots)e^{-2/z}$. For m = n, in particular, the lowest term is the constant 2^n.

LEMMA 1. *The first* n *moments of* $u_n(z)$, *with respect to the weight function* $w(z) = e^{-z/2}$, *vanish.*

Proof. $$\frac{d^n}{dz^n}(z^{2n}e^{-2/z}) = \frac{d^n}{dz^n}\sum_{\nu=0}^{\infty}(-1)^{\nu}\frac{2^{\nu}}{\nu!z^{\nu}}z^{2n} = \sum_{\nu=0}^{\infty}\frac{(-1)^{\nu}}{\nu!}2^{\nu}\frac{d^n}{dz^n}z^{-\nu+2n}$$

$$= \sum_{\nu=0}^{\infty}(-1)^{\nu}\frac{2^{\nu}}{\nu!}(-\nu+2n)(-\nu+2n-1)\ldots(-\nu+n+1)z^{-\nu+n}.$$

One observes, in particular, that the powers $-\nu+n$ corresponding to $\nu = 2n$, $\nu = 2n-1,\ldots,\nu = n+1$ do not occur, so that (see remark to that effect in [17])

$$\frac{d^n}{dz^n}(z^{2n}e^{-2/z}) = \sum_{\nu=0}^{n}(-1)^{\nu}\frac{2^{\nu}}{\nu!}(2n-\nu)(2n-1-\nu)\ldots(n+1-\nu)z^{n-\nu}$$

$$+\sum_{\nu=2n+1}^{\infty}\frac{2^{\nu}}{\nu!}(\nu-2n)(\nu-2n+1)\ldots(\nu-n-1)z^{-\nu+n}.$$

This is the sum of a polynomial of exact degree n and of an infinite series of decreasing powers, that starts with z^{-n-1}. It also follows that $f_{k,n}(z) = z^k\frac{d^n}{dz^n}(z^{2n}e^{-2/z})$ is the sum of a polynomial of exact degree k+n and of a series of decreasing powers starting with z^{k-n-1}. Consequently, for k = 0,1,...,n-1, the residues of $f_{k,n}(z)$ vanish, so that

$$\int_C e^{-2/z}z^k u_n(z)dz = \int_C e^{-2/z}z^k\{e^{2/z}\frac{d^n}{dz^n}(z^{2n}e^{-2/z})\}dz =$$

$$\int_C z^k\frac{d^n}{dz^n}(z^{2n}e^{-2/z})dz = \int_C f_{k,n}(z)dz = 0, \text{ as claimed.}$$

On the other hand, we know from Corollary 4.4 that $\int_C e^{-2/z}z^k y_n(z)dz = 0$ for k = 0,1,...,n-1. From Corollary 1 it now follows that $u_n(z)$ can differ from $y_n(z)$ only by an arbitrary multiplicative factor.

THEOREM 1. (see [68]). *The* BP $y_n(z)$ *admits the Rodrigues' formula*

$$(2'') \qquad y_n(z) = 2^{-n} e^{2/z} \frac{d^n}{dz^n} (z^{2n} e^{-2/z}).$$

Proof. On account of what precedes, it only remains to determine the factor of proportionality. We know from Proposition 4, that the constant term of $u_n(z)$ is 2^n. Also (see Chapter 2) the constant term of $y_n(z)$ is 1 and Theorem 1 is proved.

A somewhat different proof of the Lemma may be given, following [68]. It depends on the remark that all functions involved are uniform. We proceed by integrating around C, say, from +1 to +1 and use integration by parts. All integrated terms vanish, because they are singlevalued. Specifically,

$$\int_C z^k \frac{d^n}{dz^n} (z^{2n} e^{-2/z}) dz = z^k \frac{d^{n-1}}{dz^{n-1}} (z^{2n} e^{-2/z})\Big]_{+1}^{+1} - k \int_C z^{k-1} \frac{d^{n-1}}{dz^{n-1}} (z^{2n} e^{-2/z}) dz = \ldots =$$

$$(-1)^r k(k-1)\ldots(k-r+1) \int_C z^{k-r} \frac{d^{n-r}}{dz^{n-r}} (z^{2n} e^{-2/z}) dz.$$

In particular, for $0 \le k = r < n$, this becomes

$$(-1)^r r! \int_C \frac{d^{n-r}}{dz^{n-r}} (z^{2n} e^{-2/z}) dz = (-1)^r r! \frac{d^{n-r-1}}{dz^{n-r-1}} (z^{2n} e^{-2/z})\Big]_{+1}^{+1} = 0.$$

The rest of the proof of Theorem 1 may be kept unchanged.

3. THEOREM 2. *The general* BP $y_n(z;a,b)$ *has the Rodrigues' formula*

$$(2) \qquad y_n(z;a,b) = b^{-n} z^{2-a} e^{b/z} \frac{d^n}{dz^n} (z^{2n+a-2} e^{-b/z}).$$

For $a = b = 2$, (2) reduces to (2''). It also is clear that not much generality is gained by allowing arbitrary values of $b \neq 2$. One could well consider instead of (2) simply

$$(2') \qquad y_n(z;a) = 2^{-n} z^{2-a} e^{2/z} \frac{d^n}{dz^n} (z^{2n+a-2} e^{-2/z}),$$

from which (2) immediately follows by a change of variable. However, the proof of (2) is not more difficult than that of (2'), so that we proceed to prove Theorem 2 as stated.

In the proof we shall need

LEMMA 2. *For* $k \le n$, $\sum_{s=0}^{k} (-1)^s \binom{k}{s} (2n+a-2-s)^{(n)} = n^{(k)} (2n+a-k-2)^{(n-k)}$; *for* $k > n$, *the sum vanishes.*

Proof. For integers n,c,k,

$$\frac{d^n}{dx^n}\{x^c(x-1)^k\} = \frac{d^n}{dx^n}\sum_{s=0}^{k}(-1)^s\binom{k}{s}x^{c+k-s} = \sum_{s=0}^{k}(-1)^s\binom{k}{s}(c+k-s)^{(n)}x^{c+k-s-n}.$$

For $x = 1$, in particular, the second member becomes $\sum_{s=0}^{k}(-1)^s\binom{k}{s}(c+k-s)^{(n)}$. On the other hand, by Leibniz's rule, the first member equals $\sum_{r=0}^{n}\binom{n}{r}\left(\frac{d^{n-r}}{dx^{n-r}}x^c\right)\left(\frac{d^r}{dx^r}(x-1)^k\right)$.

The terms with $r > k$ vanish, because $(x-1)^k$ is of lower degree than r; for $r < k$ the last factor contains a positive power of $x-1$ and vanishes at $x = 1$; consequently, only the term with $r = k$ can be different from zero. It exists, however, only if $k \le n$ and then its value is $\binom{n}{k}c^{(n-k)}k! = n^{(k)}c^{(n-k)}$. We have proved that for any integers n,c,k, $k \le n$, $\sum_{s=0}^{k}(-1)^s\binom{k}{s}(c+k-s)^{(n)} = n^{(k)}c^{(n-k)}$ and that the sum vanishes for $k > n$. For $c = 2n+a-k-2$ this is precisely the Lemma.

<u>Proof of Theorem 2.</u> We expand the right hand side:

$$b^{-n}z^{2-a}\sum_{m=0}^{\infty}\frac{b^m}{m!z^m}\frac{d^n}{dz^n}\sum_{s=0}^{\infty}(-1)^s\frac{b^s}{s!}z^{2n+a-2-s} =$$

$$b^{-n}z^{2-a}\sum_{m=0}^{\infty}\frac{b^m}{m!}z^{-m}\sum_{s=0}^{\infty}(-1)^s\frac{b^s}{s!}(2n+a-2-s)^{(n)}z^{n+a-2-s}.$$

Set $m+s = k$; then the double sum can be written as

$$b^{-n}\sum_{k=0}^{\infty}\frac{b^k z^{n-k}}{k!}\sum_{s=0}^{k}(-1)^s\binom{k}{s}(2n+a-2-s)^{(n)}.$$

By Lemma 2, the inner sum equals $n^{(k)}(2n+a-k-2)^{(n-k)}$ for $k \le n$, zero otherwise; hence, the sum becomes

$$\sum_{k=0}^{n}\frac{b^{k-n}z^{n-k}}{k!}n^{(k)}(2n+a-k-2)^{(n-k)} = \sum_{k=0}^{n}b^{-k}z^k\binom{n}{k}(n+k+a-2)^{(k)}$$

and this is precisely $y_n(z;a,b)$, as follows from (2.28).

4. From the Rodrigues' formula (1) for $\theta_n(z;a,b)$ immediately follows the correspond ing formula for ϕ_n, namely:

$$\phi_n(z;a,b,c) = (-1)^n b^{-n}e^{(b-c)z}z^{2n+a-1}\frac{d^n}{dz^n}(z^{-n-a+1}e^{-bz}),$$

whence, in particular

$$\phi_n(z;a,b,b) = (-1)^n b^{-n}z^{2n+a-1}\frac{d^n}{dz^n}(z^{-n-a+1}e^{-bz});$$

$$\phi_n(z;a,b) = \phi_n(z;a,b,b/2) = (-1)^n b^{-n} e^{zb/2} z^{2n+a-1} \frac{d^n}{dz^n}(z^{-n-a+1}e^{-bz});$$

$$\phi_n(z;a) = \phi_n(z;a,2) = (-1)^n 2^{-n} e^z z^{2n+a-1} \frac{d^n}{dz^n}(z^{-n-a+1}e^{-2z}); \text{ and}$$

$$\phi_n(z) = \phi_n(z;2) = (-1)^n 2^{-n} e^z z^{2n+1} \frac{d^n}{dz^n}(z^{-n-1}e^{-2z}).$$

5. In Section 1 it was suggested that a natural way to obtain Theorem 1 and 2 would be to use (1), already known from Chapter 2, and the relation $y_n(z) = z^n\theta_n(z^{-1})$. In the present section we shall sketch such a proof and similar ones. We shall show that they require certain combinatorial identities and they may appear less trivial than one may have expected.

We already know (see (6.2)) that

$$(1-2zt)^{-1/2}\exp\{[1-(1-2zt)^{1/2}]/z\} = \sum_{n=0}^{\infty} y_n(z)t^n/n!\ ,$$

so that $y_n(z) = \frac{d^n}{dt^n}((1-2zt)^{-1/2}\exp\{[1-(1-2zt)^{1/2}]/z\})_{t=0}$.

As z is not involved in the differentiation,

$$\text{(3)}\qquad e^{-2/z}y_n(z) = \frac{d^n}{dt^n}((1-2zt)^{-1/2}\exp\{-[1+(1-2zt)^{1/2}]/z\})_{t=0} = n!c_n,$$

where c_n is the coefficient of t^n in the expansion of the large bracket. Let $v = (1-2zt)^{1/2}$. Then

$$\frac{1}{v}e^{-(1+v)/z} = \frac{1}{v}\sum_{k=0}^{\infty}(-1)^k\frac{(1+v)^k}{k!z^k} = \frac{1}{v}\sum_{k=0}^{\infty}\frac{(-1)^k}{k!z^k}\sum_{r=0}^{k}\binom{k}{r}v^r =$$

$$\sum_{k=0}^{\infty}\frac{(-1)^k}{k!z^k}\sum_{r=0}^{k}\binom{k}{r}(1-2tz)^{(r-1)/2} = \sum_{k=0}^{\infty}\frac{(-1)^k}{k!z^k}\sum_{r=0}^{k}\binom{k}{r}\sum_{m=0}^{\infty}(-1)^m\binom{\frac{r-1}{2}}{m}2^m z^m t^m$$

$$= \sum_{m=0}^{\infty}(-1)^m t^m.2^m z^m\sum_{k=0}^{\infty}\frac{(-1)^k}{k!z^k}\sum_{r=0}^{k}\binom{k}{r}\binom{\frac{r-1}{2}}{m} = \sum_{m=0}^{\infty}c_m t^m.$$

It follows that the coefficient c_n of t^n is

$$c_n = (-1)^n 2^n z^n\sum_{k=0}^{\infty}\frac{(-1)^k}{k!z^k}\sum_{r=0}^{k}\binom{k}{r}\binom{\frac{r-1}{2}}{n}\ ,\ \text{or, by (3)},$$

$$y_n(z) = (-1)^n e^{2/z} n!2^n z^n\sum_{k=0}^{\infty}\frac{(-1)^k}{k!z^n}\sum_{r=0}^{k}\binom{k}{r}\binom{\frac{r-1}{2}}{n}.$$

By (2") this implies the validity of the following identity:

$$(4)\qquad (-1)^n 2^n z^n \sum_{k=0}^{\infty} \frac{(-1)^k}{k!\,z^k} \sum_{r=0}^{k} \binom{k}{r} \binom{\frac{r-1}{2}}{n} = \frac{2^{-n}}{n!} \frac{d^n}{dz^n} (z^{2n} e^{-2/z}).$$

However, $\frac{d^n}{dz^n} (z^{2n} e^{-2/z}) = \sum_{k=0}^{\infty} (-1)^k \frac{2^k}{k!} \frac{d^n}{dz^n} z^{2n-k} =$

$$\sum_{k=0}^{\infty} (-1)^k \frac{2^k}{k!} (2n-k)^{(n)} z^{n-k} = \sum_{k=0}^{n} (-1)^k \frac{2^k}{k!} (2n-k)^{(n)} z^{n-k} + \sum_{k=2n+1}^{\infty} \frac{2^k}{k!} (k-2n)_{(n)} z^{-k+n},$$

because the products $(2n-k)(2n-k-1)\ldots(n-k+1)$ with $n < k \le 2n$ all vanish.

By equating the coefficients of equal powers of z in the two members of (4), it follows that (4) holds if, and only if

(a) for $k \le n$, $(-1)^n 2^{2n} n! \sum_{r=0}^{k} \binom{k}{r} \binom{\frac{r-1}{2}}{n} = 2^k (2n-k)(2n-k-1)\ldots(n-k+1)$;

(b) for $n < k \le 2n$, $\sum_{r=0}^{k} \binom{k}{r} \binom{\frac{r-1}{2}}{n} = 0$;

(c) for $k \ge 2n+1$, $2^{2n} n! \sum_{r=0}^{k} \binom{k}{r} \binom{\frac{r-1}{2}}{n} = 2^k (k-2n)(k-2n+1)\ldots(k-n-1)$.

These three conditions are jointly equivalent to

$$(5)\qquad \sum_{r=0}^{k} \binom{k}{r} \binom{\frac{r-1}{2}}{n} = 2^{k-2n} \binom{k-n-1}{n}.$$

This formula appears to be new. The closest result available in the literature seems to be Carlitz' formula (see [20])

$$(5')\qquad \sum_{r=0}^{k} \binom{k}{r} \binom{r/2}{n} \begin{cases} = \frac{k}{n} \binom{k-n-1}{n-1} 2^{k-2n} & (n \ne 0) \\ = 2^k & (n = 0) \end{cases}$$

In fact (5) and (5') can be obtained from each other, but not trivially [22]. Both are particular cases of sums of the form

$$\sum_{r=0}^{k} (\pm 1)^r \binom{k}{r} \binom{(k+s)/2}{n} \quad (a = 1 \text{ or } 2,\ s = \text{rational integer})$$

studied by H.W. Gould [21].

We have proved the purely combinatorial identity (5), valid for all integers k, by using Rodrigues' formula (2"). Conversely, however, if one knows that (5) holds (and (5) can be proved also directly, although not trivially - see [21]), then one can trace all steps backwards and one obtains the Rodrigues formula (2") as a consequence of the generating function (6.2) and (5). This is perhaps a rather artificial procedure and for this reason we abstain from proving (2), by starting similarly from the generating function (6.6).

Finally, let us see how one can obtain (2") from (1"). By (2"),

$\theta_n(z) = z^n y_n(1/z) = z^n \cdot 2^{-n} e^{2z} \{\frac{d^n}{d^n u}(u^{2n} e^{-2/u})\}_{u=1/z}$. If we also replace $\theta_n(z)$

with the help of (1") and suppress common factors, we obtain the identity

$$(6) \qquad \{\frac{d^n}{du^n}(u^{2n} e^{-2/u})\}_{u=1/z} = (-1)^n z^{n+1} \frac{d^n}{dz^n}(z^{-n-1} e^{-2z}).$$

Clearly, by going backwards, (2") is a consequence of (1") and (6), if the latter can be established directly. This we now proceed to do. The left hand member can be computed by (4) to be equal to

$$\{(-1)^n 2^{2n} n! u^n \sum_{k=0}^{\infty} \frac{(-1)^k}{k! u^k} \sum_{r=0}^{k} \binom{k}{r} \binom{\frac{r-1}{2}}{n}\}_{u=1/z} = (-1)^n 2^{2n} n! \sum_{k=0}^{\infty} \frac{(-1)^k}{k!} z^{k-n} \sum_{r=0}^{k} \binom{k}{r} \binom{\frac{r-1}{2}}{n}$$

Assuming once more that (5) holds (this was already needed in order to establish (4) directly), (6) becomes

$$(7) \qquad \frac{d^n}{dz^n}(z^{-n-1} e^{-2z}) = 2^{2n} n! \sum_{k=0}^{\infty} \frac{(-1)^k}{k!} z^{k-2n-1} 2^{k-2n} \binom{k-n-1}{n}.$$

The left hand side equals

$$\frac{d^n}{dz^n}\{z^{-n-1} \sum_{k=0}^{\infty} \frac{(-1)^k}{k!} 2^k z^k\}$$

$$= \sum_{k=0}^{\infty} \frac{(-1)^k}{k!} 2^k \frac{d^n}{dz^n} z^{k-n-1} = \sum_{k=0}^{\infty} (-1)^k \frac{2^k}{k!} (k-n-1)^{(n)} z^{k-2n-1}.$$

The coefficient of z^{k-2n-1} on the right side of (7) equals $(-1)^k \frac{n! 2^k}{k!} \frac{(k-n-1)^{(n)}}{n!}$,

so that (6) is indeed an identity. As already pointed out, (6) and (1") imply (2"). This method, however, requires the use of the combinatorial identity (5) and is hardly easier than the direct proof.

It is clear how one can prove the more general result (2), by starting from (1) instead of (1") and making use of (5), but we abstain from reproducing it here.

CHAPTER 8

THE BP AND CONTINUED FRACTIONS

1. Let $[a_0, a_1, \ldots, a_n, \ldots]$ stands for the continued fraction

$$a_0 + \cfrac{1}{a_1 + \cdots + \cfrac{1}{a_n + \cdots}}$$

of partial quotients a_n and denote the n-th convergent $[a_0, a_1, \ldots, a_n]$ by p_n/q_n, where $(p_n, q_n) = 1$. Consider, in particular, the expansion into a continued fraction of the function

$$\frac{e^{2/z} + 1}{e^{2/z} - 1} = [z, 3z, 5z, \ldots, (2n+1)z, \ldots], \tag{1}$$

known from the work of Lambert [**42**], (see also [**15**]). See [83] for a recent proof.

If we denote the numerators and denominators of its successive convergents by $p_n(z)$ and $q_n(z)$, respectively, we find, e.g.,

$$\frac{p_0(z)}{q_0(z)} = z = \frac{z}{1}\ ;\ \frac{p_1(z)}{q_1(z)} = z + \frac{1}{3z} = \frac{3z^2+1}{3z}\quad, \text{ etc.}$$

It is well known (see, e.g. [**47**]) that $p_n(z)$ and $q_n(z)$ can both be obtained, for $n \geq 2$, from the same recurrence relation

$$X_n = a_n X_{n-1} + X_{n-2}, \tag{2}$$

where one has to start with the initial values $p_0(z) = z$ and $p_1(z) = 3z^2+1$ for $p_n(z)$, and with $q_0(z) = 1$, $q_1(z) = 3z$ for $q_n(z)$. In either case $a_n = (2n+1)z$, as follows from (1). One obtains, in particular, for $n = 2$:

$$p_2(z) = p_1(z)a_2 + p_0(z) = (3z^2+1)(5z)+z = 15z^3+6z,$$

$$q_2(z) = q_1(z)a_2 + q_0(z) = (3z)(5z)+1 = 15z^2+1.$$

One may observe that (2) actually holds also for $n = 1$, provided that one defines $p_{-1}(z) = 1$, $q_{-1}(z) = 0$. One is led to make the following remarks:

(i) All powers of z that occur in $q_n(z)$ are of the same parity, which is that of n, and all powers that occur in $p_n(z)$ are of the parity opposite to that of n. This is verified directly, on above examples, for $n = 0,1,2$, and holds in general by induction on n, on account of (2).

(ii) $p_0(z) + q_0(z) = z+1 = y_1(z)$; $p_1(z) + q_1(z) = 3z^2+1+3z = y_2(z)$, and also $p_2(z)+q_2(z) = y_3(z)$. The identity $p_n(z) + q_n(z) = y_{n+1}(z)$ holds, therefore, for $n = 0,1,2$. It appears reasonable to make a change of subscripts and set $P_{n+1}(z) = p_n(z)$, $Q_{n+1}(z) = q_n(z)$, and to conjecture

THEOREM 1. $P_n(z) + Q_n(z) = y_n(z)$.

This statement holds, as seen, for $n = 1,2,3$, and, with previous convention, $p_{-1}(z) = P_0(z) = 1$, $q_{-1}(z) = Q_0(z) = 0$, it is verified to hold also for $n = 0$. The proof of Theorem 1 for all n is completed by induction on n as follows. Identity (2), written out explicitly for $P_n(z)$ and $Q_n(z)$ reads:

$$(3)\quad \begin{aligned} P_{n+1}(z) &= (2n+1)z\, P_n(z) + P_{n-1}(z) \\ Q_{n+1}(z) &= (2n+1)z\, Q_n(z) + Q_{n-1}(z). \end{aligned}$$

If we assume that Theorem 1 holds for all subscripts up to n, then (3) yields

$$\begin{aligned} P_{n+1}(z)+Q_{n+1}(z) &= (2n+1)z(P_n(z)+Q_n(z)) + (P_{n-1}(z)+Q_{n-1}(z)) \\ &= (2n+1)z\, y_n(z) + y_{n-1}(z), \end{aligned}$$

by the induction assumption. But, by (3.7), the right hand side equals $y_{n+1}(z)$, so that Theorem 1 holds also for n+1, as claimed.

In this decomposition of $y_n(z)$, $P_n(z)$ is the polynomial containing all the terms of $y_n(z)$ with powers of z of the same parity as n, while $Q_n(z)$ contains the terms with powers of z of parity opposite to that of n. In other words, $P_n(z) = \frac{1}{2}\{y_n(z)+(-1)^n y_n(-z)\}$, $Q_n(z) = \frac{1}{2}\{y_n(z)-(-1)^n y_n(-z)\}$. With this we have proved

THEOREM 2. *The* n-th *convergent of the continued fraction* $[z,3z,\ldots,(2n+1)z,\ldots]$ *of* $\dfrac{e^{2/z}+1}{e^{2/z}-1}$ *is* $\dfrac{y_n(z)+(-1)^n y_n(-z)}{y_n(z)-(-1)^n y_n(-z)}$.

2. In what follows we shall need the following

LEMMA 1. $P_{n+1}(z)Q_n(z)-P_n(z)Q_{n+1}(z) = (-1)^{n+1}$.

The lemma would follow immediately, if we knew that $P_n(z)$ and $Q_n(z)$ are coprime. This is indeed the case, both, in the algebraic sense ($P_n(z)$ and $Q_n(z)$ do not possess any non-trivial polynomial divisor) and in the arithmetic sense (for integer m, the integers $P_n(m)$ and $Q_n(m)$ are coprime), as stated, e.g. in [11], p. 83, but it is more convenient to prove the Lemma directly, without the need to invoke much classical theory of continued fractions; the coprimality will then follow as an easy corollary.

We verify that

$$P_1(z)Q_0(z)-P_0(z)Q_1(z) = p_0q_{-1}-p_{-1}q_0 = 0-1^2 = -1,$$

$$P_2(z)Q_1(z)-P_1(z)Q_2(z) = (3z^2+1)\cdot 1-z(3z) = 1,$$

$$P_3(z)Q_2(z)-P_2(z)Q_3(z) = (15z^3+6z)(3z)-(3z^2+1)(15z^2+1) = -1,$$

so that the Lemma holds for n = 0,1,2. Let us assume that it has already been verified for all subscripts up to n. One then observes that, by using (3) and the induction hypothesis one obtains:

$$P_{n+1}(z)Q_n(z)-P_n(z)Q_{n+1}(z) = [(2n+1)zP_n(z) + P_{n-1}(z)]Q_n(z)$$

$$- P_n(z)[(2n+1)zQ_n(z)+Q_{n-1}(z)] = P_{n-1}(z)Q_n(z)-P_n(z)Q_{n-1}(z)$$

$$= - [P_n(z)Q_{n-1}(z)-P_{n-1}(z)Q_n(z)] = (-1)^{n+1}$$

and the Lemma is proved.

COROLLARY 1. *For all integers* n *and* k, $(P_n(k), Q_n(k)) = 1$.

Proof. If $d = (P_n(k)), Q_n(k))$, then d divides the left hand side of the identity of Lemma 1; hence $d|(-1)^{n+1}$, so that d = 1, as claimed.

COROLLARY 2.

$$\frac{P_n(z)}{Q_n(z)} - \frac{P_{n+1}(z)}{Q_{n+1}(z)} = \frac{(-1)^n}{Q_n(z)Q_{n+1}(z)}$$

holds for all $z \in \mathbb{C}$, *such that* $Q_n(z)Q_{n+1}(z) \neq 0$.

Proof follows trivially from the Lemma.

We recall (see [47]) that any real number falls between any two consecutive of its convergents. More precisely, we have

COROLLARY 3. *If* n *is even and* z = x *is real and positive, then*

$$\frac{P_{n+1}(z)}{Q_{n+1}(z)} < \frac{e^{2/x}+1}{e^{2/x}-1} < \frac{P_n(x)}{Q_n(x)};$$

if n *is odd, the inequalities are reversed.*

Proof follows from the preceding remark and Corollary 2, by observing that $P_n(x)/Q_n(x)$ is the n-th convergent of $\frac{e^{2/x}+1}{e^{2/x}-1}$ and that $Q_n(x)Q_{n+1}(x) > 0$ for $x > 0$

COROLLARY 4. *Regardless of the parity of* n *and for all real* x *such that* $Q_n(x)Q_{n+1}(x) \neq 0$,

$$0 < \left| \frac{P_n(x)}{Q_n(x)} - \frac{e^{2/x}+1}{e^{2/x}-1} \right| < \frac{1}{Q_n(x)Q_{n+1}(x)}.$$

Proof. The condition $Q_n(x)Q_{n+1}(x) \neq 0$ implies $x \neq 0$. The first inequality then follows from the fact that $P_n(x)/Q_n(x)$ is a convergent with following partial quotient different from zero (in fact, by (1) a non-vanishing multiple of x); while the second inequality follows from Corollaries 2 and 3.

We now observe that $\frac{e^{2/x}+1}{e^{2/x}-1} - 1 = \frac{2}{e^{2/x}-1}$ and

$$\frac{P_n(x)}{Q_n(x)} - 1 = \frac{y_n(x)+(-1)^n y_n(-x)}{y_n(x)-(-1)^n y_n(-x)} - 1 = \frac{(-1)^n y_n(-x)}{Q_n(x)},$$ so that by Corollary 4,

$$\left| \frac{2}{e^{2/x}-1} - \frac{(-1)^n y_n(-x)}{Q_n(x)} \right| < \frac{1}{Q_n(x)Q_{n+1}(x)},$$ or, equivalently,

$$|2\, Q_n(x) - (-1)^n y_n(-x)(e^{2/x}-1)| < (e^{2/x}-1)/Q_{n+1}(x).$$

If we replace here x by 1/x and then multiply by x^n we obtain

$$(4) \qquad |2x^n Q_n(1/x) - (-1)^n x^n y_n(-1/x)(e^{2x}-1)| < (e^{2x}-1)x^{2n}/x^n Q_{n+1}(1/x).$$

In the first member, $x^n Q_n(1/x) = \tilde{Q}_n(x)$, say, collects the terms with odd powers of $\theta_n(x)$, while $(-1)^n x^n y_n(-1/x) = (-x)^n y_n(1/(-x)) = \theta_n(-x)$ and $x^n Q_{n+1}(1/x) = x^{-1}\tilde{Q}_{n+1}(x)$, a polynomial with only even powers of x. The first member of (4) may now be written as $|2\,\tilde{Q}_n(x) - \theta_n(-x)(e^{2x}-1)| = |2\tilde{Q}_n(x) + \theta_n(-x) - \theta_n(-x)e^{2x}|$. But $2\tilde{Q}_n(x) = \theta_n(x) - \theta_n(-x)$ so that $2\tilde{Q}(x) + \theta_n(-x) = \theta_n(x)$. Next, by a previous remark, $x^n Q_{n+1}(1/x) = \sum_{j=0}^{[n/2]} c_j x^{2j}$, so that $x^{2n}(e^{2x}-1)/x^n Q_{n+1}(1/x) =$

$x^{2n} \sum_{m=1}^{\infty} \frac{(2x)^m}{m!} \Big/ \sum_{j=0}^{[n/2]} c_j x^{2j}$. Here $c_0 = \frac{(2n+2)!}{2^{n+1}(n+1)!}$ as follows from Chapter 2.

This value is here without further relevance, except for $c_0 \neq 0$. If $|x|$ is less than the smallest zero of the denominator, it follows that the right hand side of (4) can be written as a series $x^{2n+1} \sum_{m=0}^{\infty} k_m x^m$ and has a zero of order 2n+1.

Consequently, also the entire function $\tilde{R}_n(x) = \theta_n(x) - \theta_n(-x)e^{2x}$, which, except perhaps for sign, is the left hand side of (4), has a zero of order at least 2n+1 at $x = 0$. Let us replace here x by x/2 and set $\tilde{R}_n(x/2) = R_n(x)$. The result obtained so far may be written as

$$\theta_n(x/2) - \theta_n(-x/2)e^x = R_n(x) = x^{2n+1} \sum_{j=0}^{\infty} c_j' x^j. \tag{6}$$

Finally, let us change notations and set $A_n(x) = \theta_n(x/2)$, $B_n(x) = -\theta_n(-x/2)$. Then (6) reads:

$$R_n(x) = A_n(x) + B_n(x)e^x = x^{2n+1} \sum_{j=0}^{\infty} c_j' x^j.$$

This is precisely the expression encountered in Problem 1 of the Introduction and shows that indeed, except for normalization, the polynomials $A_n(x)$ and $B_n(x)$ there defined, are BP. In fact, a count of the number of constants involved shows that, except for an arbitrary multiplicative constant, $A_n(x) = \theta_n(x/2)$ and $B_n(x) = -\theta_n(-x/2)$ are the only polynomials of a degree not in excess of n, and for which the linear form $R_n(x)$ has a zero of order $\geq$ 2n+1 at the origin.

CHAPTER 9

EXPANSIONS OF FUNCTIONS IN SERIES OF BP

1. The topic of expansions of arbitrary functions either of a complex, or of a real variable in series of BP has not been explored fully and many open problems remain to be answered. To date the most comprehensive treatment is that of R.P. Boas, Jr. and R.C. Buck [13], but the literature available on this subject cannot be compared to that on expansion in, say, Fourier series, or in series of classical orthonormal polynomials, etc. In view of the orthogonality of the BP on the unit circle, as presented in Chapter 4, it is reasonable to expect that the natural problem to investigate is that of the expansion of functions f(z), defined for a complex variable z, and continuous, or at least integrable on a set containing the unit circle, or some compact neighborhood of the origin.

On the other hand, one may ask just for some formal expansions, say

$f(z) \sim \sum_{n=0}^{\infty} \tilde{c}_n y_n(z;a,b)$, or $f(z) \sim \sum_{n=0}^{\infty} c_n \theta_n(z;a,b)$, and then inquire in what sense the series "represent" f(z), under what conditions they converge and, if they do, when do they converge to f(z). Even if the series do not converge, it may still be possible to find summation methods, under which the series are summable to f(z). Finally, one may raise questions concerning the uniqueness of such expansions.

The question of formal expansions into series of the BP $y_n(z;a,b)$ has been asked and answered already by Krall and Frink [68]. By using the orthogonality relations of Chapter 4, it follows that, if $f(z) = \sum_{m=0}^{\infty} \tilde{c}_m y_n(z;a,b)$, then

$$\int_\Gamma f(z) y_n(z;a,b)\rho(z)dz = \int_\Gamma \Big(\sum_{m=0}^{\infty} \tilde{c}_m y_m(z;a,b)\Big) y_n(z;a,b)\rho(z)dz,$$

where Γ is any simply closed curve that encircles the origin, and $\rho(z) = (2\pi i)^{-1} e^{-2/z}$ if a = b = 2; otherwise $\rho(z) = \rho(z;a,b)$, as defined in Chapter 4.

If Γ belongs to a connected domain on which the convergence of the series $\sum_{m=0}^{\infty} \tilde{c}_m y_m(z;a,b)$ is sufficiently strong to justify the inversion of the order of summation and integration, the last integral is equal to

$$\sum_{m=0}^{\infty} \tilde{c}_m \int_\Gamma y_m(z;a,b) y_n(z;a,b)\rho(z)dz = \tilde{c}_n \frac{(-1)^{n+1} b.n!\,\Gamma(a)}{(2n+a-1)\Gamma(n+a-1)},$$

by Corollary 4.6. In the particular case a = b = 2, this simplifies to (see [68])

$$\frac{1}{2\pi i} \int_\Gamma f(z) y_n(z) e^{-2/z} dz = (-1)^{n+1} \frac{2}{2n+1} \tilde{c}_n,$$

or

$$(1) \qquad \tilde{c}_n = (-1)^{n+1} \frac{n + 1/2}{2\pi i} \int_\Gamma f(z) y_n(z) e^{-2/z} dz.$$

If $f(z) = \sum_{m=0}^{\infty} b_m z^m$, $b_m = f^{(m)}(0)/m!$, then (1) becomes (see [81])

$$\tilde{c}_n = (-1)^{n+1}(n+\tfrac{1}{2}) \sum_{m=0}^{\infty} \frac{f^{(m)}(0)}{m!} \{\frac{1}{2\pi i} \int_\Gamma z^m y_n(z) e^{-2/z} dz\}.$$

If we replace here the curly bracket by its value from Corollary 4.4, we obtain

$$\tilde{c}_n = (-1)^{n+1}(n+\tfrac{1}{2}) \sum_{m=0}^{\infty} \frac{f^{(m)}(0)}{m!} (-2)^{m+1} \frac{m!}{(m+n+1)!(m-n)!}$$

$$= (-1)^n(2n+1) \sum_{m=0}^{\infty} \frac{f^{(m)}(0)(-2)^m}{(m+n+1)!(m-n)!}.$$

In the sum all terms with $m < n$ vanish; in the others we set $m = n + \nu$, $\nu = 0,1,\ldots$ and obtain Nasif's [81] formula:

$$\tilde{c}_n = (2n+1)\cdot 2^n \sum_{\nu=0}^{\infty} \frac{f^{(n+\nu)}(0)}{\nu!} \frac{(-2)^\nu}{(2n+\nu+1)!}$$

This, of course, is all formal work. In general, there is no guarantee that the series $\sum_{n=0}^{\infty} \tilde{c}_n y_n(z;a,b)$ converges, or even if it converges, that its sum is $f(z)$. The determination of necessary and sufficient conditions on $f(z)$, for the convergence, or the summability to $f(z)$ of the series appears to be an open problem.

2. Similar considerations hold if we consider expansions of functions $f(z)$ into series $\sum_{n=0}^{\infty} c_n \theta_n(z;a,b)$. It is clear, however, from the work of Burchnall (see Corollary 4.15) that the $\theta_n(z;a,b)$ themselves do not form a system of polynomials orthogonal on the unit circle, because the factor $z^{-(m+n+a)}e^{-bz}$ is not independent of the polynomials θ_m and θ_n involved. A somewhat different approach, however, permits one to obtain expansions of entire functions in series of the polynomials $\theta_n(z;a,b)$. It turns out that we obtain particularly simple results, if we set, at least provisionally, $p_n(z)$ $(=p_n(z;a))$ $= \frac{2^n}{n!} \theta_n(z;a,2)$. In this case, the theory of generalized Appell Polynomials, due to Boas and Buck [13], [5], permits one to obtain not only the coefficients c_n of the formal expansions, but also to determine the number

of distinct expansions (the expansions, unfortunately, are in general not unique), their respective regions of convergence and the regions of summability for these series of generalized Appell Polynomials. The main results and sketches of the proofs will be given here, but for detailed proofs it is suggested that the reader consult the original presentations [5] and [13].

3. The fundamental ideas may be traced back, on the one hand to Whittaker [66], on the other hand to the classical work of E. Borel, G. Pólya, Phragmén and Lindelöf (see [32], vol.2 and [4]).

Let $A(w) = \sum_{n=0}^{\infty} a_n w^n$, $a_0 \neq 0$; and $g(w) = \sum_{n=1}^{\infty} g_n w^n$, $g_1 \neq 0$ be given functions, holomorphic in some neighborhood of the origin. Let Ω be the largest simply connected region containing the origin, in which $A(w)$ is holomorphic, and let $\Omega_w \subseteq \Omega$ be the largest simply connected subregion of Ω, in which $\zeta = g(w)$ is univalent. Define also Δ_w as the largest disk centered at the origin and such that $\Delta_w \subseteq \Omega_w \subseteq \Omega$. Due to the univalence of $g(w)$ we have a 1:1 correspondence between the points $w \in \Omega_w$ and the images $\zeta = g(w)$. In particular, let Ω_ζ be the image of Ω_w and Δ_ζ be the image of Δ_w, so that $\Delta_\zeta \subseteq \Omega_\zeta$ in the ζ-plane.

To each entire function $f(z) = \sum_{n=0}^{\infty} f_n z^n$ we associate its Borel (or Laplace) transform $F(w) = \sum_{n=0}^{\infty} \frac{n!f_n}{w^{n+1}}$ (see [10], p. 113), holomorphic outside some circle $|z| = r$. We recall the inversion formula:

(2) $$\frac{1}{2\pi i} \int_C e^{zw} F(w)dw = \frac{1}{2\pi i} \int_C e^{zw} \sum_{n=0}^{\infty} \frac{n!f_n}{w^{n+1}} dw =$$

$$\frac{1}{2\pi i} \int_C \Big(\sum_{m=0}^{\infty} \frac{z^m w^m}{m!}\Big)\Big(\sum_{n=0}^{\infty} \frac{n!f_n}{w^{n+1}}\Big)dw = \sum_{m=0}^{\infty} \frac{z^m}{m!} \sum_{n=0}^{\infty} n!f_n \cdot \frac{1}{2\pi i} \int_C \frac{dw}{w^{n-m+1}}$$

$$= \sum_{n=0}^{\infty} f_n z^n = f(z),$$

where C is a circle of radius $r + \varepsilon$.

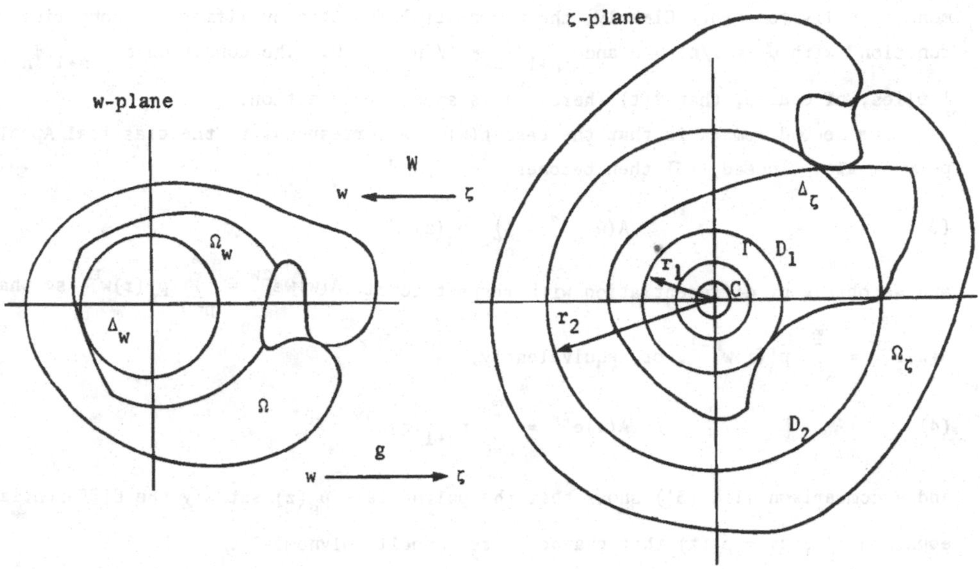

Figure 1

We shall be concerned mainly with entire functions of exponential type, i.e., functions f(z), such that $|f(z)| \le Me^{\sigma|z|}$ for finite, positive constants M, σ and all complex z. If $\tau = \inf \sigma$ for which this inequality holds, f(z) is said to be of <u>exponential type</u> τ. By D(f) we denote the complement of the set of regularity of the Borel transform F(w) of f(z).

We now define generalized Appell Polynomials $p_n(z)$ by the generating function

$$(3) \qquad A(w)e^{zg(w)} = \sum_{n=0}^{\infty} p_n(z)w^n.$$

Two remarks are in order. The first is that the theory has been developed by Boas and Buck [13] in a more general framework, in which the exponential function e^t is replaced by any "comparison function" $\Psi(t) = \sum_{n=0}^{\infty} \psi_n t^n$, subject only to the restrictions that none of the coefficients ψ_n vanish and that the ratio ψ_{n+1}/ψ_n decreases monotonically to zero. Clearly, the exponential function qualifies as "comparison function" with $\psi_n = 1/n! \neq 0$ and $\psi_{n+1}/\psi_n = 1/(n+1) \to 0$. The condition on ψ_{n+1}/ψ_n implies, of course, that $\Psi(t)$ (here e^t) is an entire function.

The second remark is that the case $g(w) = w$ corresponds to the classical Appell polynomials. Indeed, (3) then becomes

$$A(w)e^{zw} = \sum_{n=0}^{\infty} p_n(z)w^n \tag{3'}$$

and we obtain by differentiation with respect to z: $A(w)we^{zw} = \sum_{n=0}^{\infty} p_n'(z)w^n$, so that $A(w)e^{zw} = \sum_{n=1}^{\infty} p_n'(z)w^{n-1}$, or, equivalently,

$$A(w)e^{zw} = \sum_{n=0}^{\infty} p_{n+1}'(z)w^n \tag{4}$$

and a comparison with (3') shows that the polynomials $p_n(z)$ satisfy the differential equation $p_{n+1}'(z) = p_n(z)$ that characterizes Appell polynomials.

We now return to the general case. By use of the univalent map $\zeta = g(w)$, we can reduce the general case to a more manageable one, close to the classical (4). Indeed, let $w = W(\zeta)$ be the map inverse to $\zeta = g(w)$ and set $A(w) = A(W(\zeta)) = B(\zeta)$. To any compact set $C \subset \Delta_\zeta$ corresponds a class of entire functions $f(z)$, such that $D(f) \subset C$; let us denote that class by $K[C]$. Let Γ be a simply closed curve in Δ_ζ, enclosing C and passing through no zeros of $B(\zeta)$. Now (3) becomes:

$$B(\zeta)e^{\zeta z} = \sum_{n=0}^{\infty} p_n(z)W(\zeta)^n, \tag{5}$$

and it follows that on Γ, $\sum_{n=0}^{\infty} p_n(z)W(\zeta)^n/B(\zeta)$ converges uniformly to $e^{\zeta z}$.

We now set

$$c_n = L_n(f) = \frac{1}{2\pi i} \int_\Gamma \frac{W(\zeta)^n}{B(\zeta)} F(\zeta)d\zeta \tag{6}$$

and obtain, using (2), that

$$\sum_{n=0}^{\infty} c_n p_n(z) = \frac{1}{2\pi i} \int_\Gamma \frac{F(\zeta)}{B(\zeta)} \left(\sum_{n=0}^{\infty} p_n W(\zeta)^n\right)d\zeta = \frac{1}{2\pi i} \int_\Gamma F(\zeta)e^{z\zeta}d\zeta = f(z).$$

The interchange of summation and integration is justified for $f \in K[C]$, by the uniform convergence of $\sum_{n=0}^{\infty} p_n(z)W(\zeta)^n$.

In order to have $f \in K[C]$ we need to know that $D(f) \subset C \subseteq \Delta_\zeta$. In particular, if r_1 is the radius of D_1, the largest disk, centered at the origin and inside Δ_ζ, a sufficient condition for $D(f) \subset D_1 \subseteq \Delta_\zeta$ is that $\tau < r_1$. We consider also D_2, the largest disk centered at the origin, such that $D_2 \subseteq \Omega_\zeta$; let r_2 be the radius of D_2. The following theorem holds and most of its assertions follow readily from the preceding discussion.

THEOREM 1. *Let* $A(w) = \sum_{n=0}^{\infty} a_n w^n$, $a_0 \neq 0$ *and* $g(w) = \sum_{n=1}^{\infty} g_n w^n$, $g_1 \neq 0$ *be holomorphic in some neighborhood of the origin, with* $g(w)$ *univalent on* Ω_w; *let* $W(\zeta)$ *be the function inverse to* $g(w)$, *defined on* Ω_ζ , *the image of* Ω_w, *set* $A(W(\zeta)) = B(\zeta)$ *and define the polynomials* $p_n(z)$ *by* (5).

Let $f(z) = \sum_{n=0}^{\infty} f_n z^n$ *be an entire function of exponential type with Borel-Laplace transform* $F(\zeta)$, *and define the constants* c_n *by* (6). *Then* $f(z)$ *is represented by the series* $f(z) \sim \sum_{n=0}^{\infty} c_n p_n(z)$. *Furthermore, if* D_1, r_1 *and* D_2, r_2 *are defined as above and if the type* τ *of* $f(z)$ *satisfies* $\tau < r_1$, *then* $f \in K[C]$ *for some* $C \supset D(f)$ *and the series converges to* $f(z)$. *If* $\tau < r_2$, *then the series is Mittag-Leffler summable to* $f(z)$.

REMARK 1. These conditions insure convergence, or summability for all entire functions of exponential type τ, bounded by r_1, or by r_2, respectively. However, what one needs is only the existence of simply closed curves Γ, that encircle $D(f)$. Hence, for specific functions, for which $D(f)$ is explicitly known, one can often establish convergence, or summability, when the sufficient conditions $\tau < r_1$, or $\tau < r_2$ do not hold.

REMARK 2. We abstain here from a proof of the Mittag-Leffler summability in the case $r_1 < \tau < r_2$, for which the reader may want to consult [13].

We observe that $F(\zeta)$ has an essential singularity at the origin (unless $f(z)$ is a polynomial, in which case $\tau = 0$ and $F(\zeta)$ has a pole at the origin), so that the c_n do not all vanish. $B(\zeta)$ may, or may not have zeros. If it has no zeros, then any Γ selected as before, will lead to the same values for $c_n = L_n(f)$ and the series obtained coincides with Whittaker's basic series [**66**]. If, however, $B(\zeta)$ has zeros in Δ_ζ, outside $D(f)$, then the values of the c_n will, in general, depend on the set of zeros of $B(\zeta)$ enclosed by Γ, so that the expansion of $f(z)$ need not be unique. Further different expansions may be obtained, if one selects Γ outside Ω_ζ (so that

$g(w) = \zeta$ is no longer univalent), but, fortunately, these complexities need not worry us in the present case.

4. We start from Burchnall's generating function (6.6). As usually we shall set $b = 2$ (otherwise, we make the change of variables $z'/2 = z/b$) and, writing also $2z$ for z, obtain

$$\{\frac{1}{2}[1+(1-4zt)^{1/2}]\}^{2-a}(1-4zt)^{-1/2}\exp\{[1-(1-4zt)^{1/2}]/2z\} = \sum_{n=0}^{\infty} \frac{y_n(2z;a)}{n!} t^n.$$

The polynomials $y_n(2z;a)(=y_n(2z;a,2))$ as here defined, do not appear to be (generalized) Appell polynomials, according to the definition (3). Let us replace, however, t by w/z; we obtain

$$\{\frac{1}{2}[1+(1-4w)^{1/2}\}^{2-a}(1-4w)^{-1/4}\exp\{[1-(1-4w)^{1/2}]/2z\} = \sum_{n=0}^{\infty} \frac{1}{z^n} \frac{y_n(2z;a)}{n!} w^n$$

$$= \sum_{n=0}^{\infty} \frac{2^n}{n!} \frac{1}{(2z)^n} y_n(2z;a)w^n = \sum_{n=0}^{\infty} \frac{2^n}{n!} \theta_n((2z)^{-1};a)w^n,$$

or, replacing $1/2z$ by z,

$$\{\frac{1}{2}[1+(1-4w)^{1/2}\}^{2-a}(1-4w)^{-1/4}\exp\{z[1-(1-4w)^{1/2}]\} = \sum_{n=0}^{\infty} \frac{2^n}{n!} \theta_n(z;a)w^n.$$

This is precisely (3), with $A(w) = \frac{1}{2}[1+(1-4w)^{1/2}]^{2-a}(1-4w)^{-1/4}$, $g(w) = 1-(1-4w)^{1/2}$, and $p_n(z) = \frac{2^n}{n!}\theta_n(z;a)$.

In order to insure that $A(w)$ and $g(w)$ are holomorphic, it is sufficient to insure the singlevaluedness of $(1-4w)^{1/2}$. A convenient way to obtain this is to cut the complex w-plane along the positive real axis from $+1/4$ to ∞. One easily verifies that $g(w)$ is also univalent in the cut plane; hence, the cut plane is Ω_w and Δ_w is the disk $|w| < 1/4$.

The image Ω_ζ of Ω_w under $\zeta = g(w) = 1-(1-4w)^{1/2}$ is found, by observing that the image of the boundary $\frac{1}{2} < w < \infty$ of Ω_w becomes $1 \mp i(4w-1)^{1/2} = 1 \mp iv$ $(0 < v < \infty)$ in the ζ-plane. It follows that Ω_ζ is the half-plane $\mathrm{Re}\,\zeta < 1$. The function $w = W(\zeta)$, inverse to $g(w)$, is obtained from $(1-4w)^{1/2} = 1-\zeta$, $1-4w = 1-2\zeta+\zeta^2$, $w = \frac{1}{4}\zeta(2-\zeta)$. The image of Δ_w, i.e., of the disk $|w| < \frac{1}{4}$ is, therefore, that part of the inside of the lemniscate $|\zeta(2-\zeta)| < 1$, that is contained in Ω_ζ, i.e., the left loop of it. The lemniscate is symmetric with respect to the x-axis, which it meets in the points $\zeta^2-2\zeta\pm 1 = 0$. For the plus sign we obtain the double point

$\zeta = 1$ of the lemniscate, on the boundary of Ω_ζ; for the minus sign we obtain the vertices of the lemniscate, at $\zeta = 1 \pm \sqrt{2}$. The first one, $\zeta = 1 + \sqrt{2}$, is not in Ω_ζ, while $\zeta = 1-\sqrt{2} = -(\sqrt{2}-1)$, is in Ω_ζ. The largest disk D_1 inside Δ_ζ has, therefor, a radius $r_1 = \sqrt{2}-1$. This requires some verification, in particular, that no point of the lemniscate is closer to the origin than this vertex at $-(\sqrt{2}-1)$.

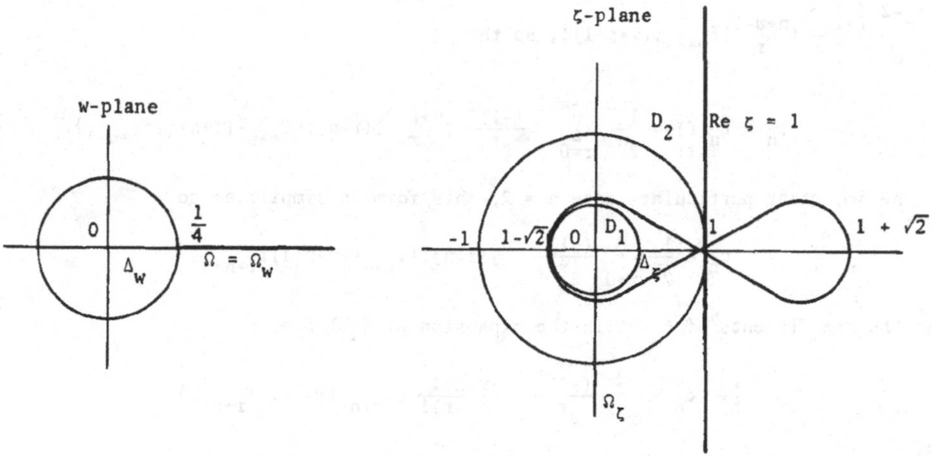

Figure 2

The disk D_2 has the radius $r_2 = 1$. A simple computation shows that $B(\zeta) = A(W(\zeta)) = \{\frac{1}{2}[1+1-\zeta]\}^{2-a}(1-\zeta)^{-1} = (1-\zeta/2)^{2-a}/(1-\zeta)$ is singlevalued and does not vanish in Ω_ζ, where $\operatorname{Re}\zeta < 1$. It follows that the series $\sum_{n=0}^{\infty} \frac{2^n}{n!} c_n \theta_n(z;a)$ has uniquely defined coefficients c_n. For $f(z) = \sum_{n=0}^{\infty} f_n z^n$, $L_n(f)$ becomes successively

$$\frac{1}{2\pi i}\int_\Gamma \frac{1}{4^n}\frac{\zeta^n(2-\zeta)^n(1-\zeta)}{(1-\zeta/2)^{2-a}}F(\zeta)d\zeta = \frac{1}{2^n}\frac{1}{2\pi i}\int_\Gamma \zeta^n(1-\zeta)(1-\zeta/2)^{n+a-2}F(\zeta)d\zeta =$$

$$\frac{1}{2^n}\frac{1}{2\pi i}\int_\Gamma \zeta^n(1-\zeta)(1-\zeta/2)^{n+a-2}\Big(\sum_{m=0}^{\infty}\frac{m!f_m}{\zeta^{m+1}}\Big)d\zeta =$$

$$\frac{1}{2^n}\sum_{m=0}^{\infty} m!\, f_m \frac{1}{2\pi i}\int_\Gamma \frac{(1-\zeta)(1-\zeta/2)^{n+a-2}}{\zeta^{m-n+1}}\, d\zeta.$$

In particular, for integral $a \geq 1$, the numerator equals $\sum_{r=0}^{n+a-2} \frac{(-1)^r}{2^r} \binom{n+a-2}{r} (\zeta^r - \zeta^{r+1})$ and the integrand becomes $\sum_{r=0}^{n+a-2} \frac{(-1)^r}{2^r} \binom{n+a-2}{r} \{\frac{1}{\zeta^{m-r-n+1}} - \frac{1}{\zeta^{m-r-n}}\}$. The integrals vanish, except for $m = r+n$ and $m = r+n+1$. The first case contributes $\sum_{r=0}^{n+a-2} \frac{(-1)^r}{2^r} \binom{n+a-2}{r} f_{r+n}(r+n)!$, while the second contributes $-\sum_{r=0}^{n+a-2} \frac{(-1)^r}{2^r} \binom{n+a-2}{r} f_{r+n+1}(r+n+1)!$, so that

$$c_n = L_n(f) = \frac{1}{2^n} \sum_{r=0}^{n+a-2} \frac{(-1)^r}{2^r} \binom{n+a-2}{r} (r+n)!\{f_{r+n} - (r+n+1)f_{r+n+1}\}. \tag{7}$$

In the important particular case $a = 2$, this formula simplifies to

$$c_n = \frac{1}{2^n} \sum_{r=0}^{n} \frac{(-1)^r}{2^r} \binom{n}{r} (r+n)!\{f_{r+n} - (r+n+1)f_{r+n+1}\}$$

and the coefficients of $\theta_n(z)$ in the expansion of $f(z)$ are

$$\begin{aligned} \frac{2^n}{n!} c_n &= \sum_{r=0}^{n} \frac{(-1)^r}{2^r} \frac{(r+n)!}{r!(n-r)!} \{f_{r+n} - (r+n+1)f_{r+n+1}\} \\ &= \sum_{r=0}^{n} (-1)^r a_r^{(n)} (f_{r+n} - (r+n+1)f_{r+n+1}), \end{aligned} \tag{8}$$

with the $a_r^{(n)}$ given by (2.8), as coefficients of $y_n(z) = \sum_{r=0}^{n} a_r^{(n)} z^r$.

On account of the fact that $r_1 = \sqrt{2} - 1$ and $r_2 = 1$, it follows that, as long as a is a natural integer, the expansion $f(z) = \sum_{n=0}^{\infty} \frac{2^n}{n!} c_n \theta_n(z;a)$ with the c_n given by (7) are convergent and converge to $f(z)$, at least for entire functions $f(z)$ of exponential type $\tau < \sqrt{2} - 1$. If $\tau < 1$, then the series is at least Mittag-Leffler summable to $f(z)$.

5. As an application, let us find the expansions of powers of z in series of BP. If $f(z) = z^k$, then $f_k = 1$, $f_r = 0$ for $r \neq k$, so that, by (8),

$$z^k = \sum_{\substack{n+r=k \\ 0 \leq r \leq n}} \frac{(-1)^r}{2^r} \frac{k!}{r!(n-r)!} \theta_n(z) - \sum_{\substack{n+r+1=k \\ 0 \leq r \leq n}} \frac{(-1)^r}{2^r} \frac{k!}{r!(n-r)!} \theta_n(z).$$

In the first sum $r \leq n \leq k-r$, $0 \leq n-r \leq k-2r$ and the sum extends over $0 \leq r \leq k/2$;

in the second sum, similarly, $0 \le n-r \le k-2r-1$ and the sum extends over $0 \le r \le (k-1)/2$ Consequently,

$$z^k = k! \sum_{r=0}^{k/2} \frac{(-1)^r}{2^r} \frac{1}{r!(k-2r)!} \theta_{k-r}(z) - k! \sum_{r=0}^{(k-1)/2} \frac{(-1)^r}{2^r} \frac{1}{r!(k-2r-1)!} \theta_{k-r-1}(z).$$

In the second sum replace r+1 by r, which will then run from 1 to (k+1)/2 and reads, including sign, $\sum_{r=1}^{(k+1)/2} \frac{(-1)^r}{2^{r-1}} \frac{1}{(r-1)!(k-2r+1)!} \theta_{k-r}(z)$

$= \sum_{r=1}^{(k+1)/2} \frac{(-1)^r \cdot 2r}{2^r r!(k-2r+1)!} \theta_{k-r}(z)$. For $1 \le r \le k/2$ we can combine the two sums.

The general term is

$$k! \frac{(-1)^r}{2^r} \frac{(k-2r+1)+2r}{r!(k-2r+1)!} \theta_{k-r}(z) = (-1)^r \frac{(k+1)!}{2^r r!(k-2r+1)!} \theta_{k-r}(z).$$

Finally, we verify that for r = 0 and r = (k+1)/2 this general term represents precisely the omitted terms, the one for r = 0 of the first sum, and the one for r = (k+1)/2 of the last sum, respectively, so that we obtain the final result:

$$z^k = (k+1)! \sum_{r=0}^{(k+1)/2} (-1)^r \frac{\theta_{k-r}(z)}{2^r r!(k-2r+1)!} \tag{9}$$

found by Carlitz [19]. The corresponding representation of z^k by sums of the polynomials $y_n(z)$ is (in a somewhat different notation) in Dickinson [47] and Al-Salam [3] and reads

$$z^k = 2^k k! \sum_{r=0}^{k} (-1)^r \frac{2r+1}{(k-r)!(k+r+1)!} y_r(z).$$

This formula may be proved by using (1) together with Corollary 4.4.

Let us verify (9) for instance in the particular case k = 3:

$$z^3 = 4!\{\frac{1}{4!} \theta_3(z) - \frac{1}{2\cdot 1!2!} \theta_2(z) + \frac{1}{2^2 2!0!} \theta_1(z)\}$$

$$= \theta_3(z) - 6\theta_2(z) + 3\theta_1(z).$$

If we substitute for $\theta_j(z)$ the corresponding polynomials (see Chapter 3) we verify that indeed $(z^3+6z^2+15z+15) - 6(z^2+3z+3)+3(z+1) = z^3$.

6. As another application, let us find the expansion of the exponential function $f(z) = e^{\alpha z} = \sum_{k=0}^{\infty} f_k z^k$. Clearly, $f_k = \frac{\alpha^k}{k!}$ and, by (8),

$$\frac{2^n}{n!} c_n = \sum_{r=0}^{n} \frac{(-1)^r}{2^r} \frac{(r+n)!}{r!(n-r)!} \{\frac{\alpha^{r+n}}{(r+n)!} - \frac{\alpha^{r+n+1}(r+n+1)}{(r+n+1)!}\}$$

$$= \sum_{r=0}^{n} \frac{(-1)^r}{2^r} \frac{\alpha^{r+n}(1-\alpha)}{r!(n-r)!} = \frac{(1-\alpha)\alpha^n}{n!} \sum_{r=0}^{n} (-1)^r \binom{n}{r} \left(\frac{\alpha}{2}\right)^r =$$

$$(1-\alpha)(1-\alpha/2)^n \alpha^n/n! \ .$$

It follows that the expansion of $e^{\alpha z}$ in a series of BP $\theta_n(z)$ is:

$$e^{\alpha z} - (1-\alpha) \sum_{n=0}^{\infty} (\alpha-\alpha^2/2)^n \theta_n(z)/n! \ . \tag{10}$$

One easily verifies that (10) is precisely the result that one obtains, if one differentiates Carlitz' formula(6.17)with respect to u and then sets $2u = \alpha$.

As for convergence, it is clear that the right hand side of (10) converges for all complex values of α and z. Also, we recall that the Mittag-Leffler summability method is regular, so that, for convergent series, it leads to the same value as the direct summation of the series. Hence, for $|\alpha| = \tau < 1$, the series (10) converges to the function $e^{\alpha z}$. What happens for $|\alpha| \geq 1$? One observes that for $\alpha = 1$, the right hand side of (10) vanishes, while the function e^z does not. Hence, although the series converges for all α, for $|\alpha| \geq 1$ it does not converge to the function $e^{\alpha z}$. One may observe, for instance, that for $\alpha = 2$, the sum of the series equals -1, which is different, in general, from $f(z) = e^{2z}$.

If we set $z = 0$ in (10) and replace $\theta_n(0)$ by its value $(2n)!/2^n n!$, we obtain after a short computation that, for $|\alpha| < 1/2$,

$$\sum_{n=0}^{\infty} \binom{2n}{n} (\alpha-\alpha^2)^n = \frac{1}{1-2\alpha} = \sum_{m=0}^{\infty} 2^m \alpha^m. \tag{11}$$

From (11) one can obtain many combinatorial identities. In particular, by comparing the coefficients of α^k, it follows that

$$\sum_{r=0}^{k/2} (-1)^r \binom{2k-2r}{k-r} \binom{k-r}{r} = 2^k.$$

PART III

CHAPTER 10

PROPERTIES OF THE ZEROS OF BP

1. In this chapter we shall be concerned mainly with properties of the zeros of the simple BP, either $y_n(z)$, or $\theta_n(z)$, which we denote by $\alpha_k^{(n)}$ and $\beta_k^{(n)}$ $(k = 1,2,\ldots,n)$, respectively. Whenever possible (and convenient) we shall extend the results to the zeros of $y_n(z;a)$, or $\theta_n(z;a)$, denoted by $\alpha_k^{(n)}(a)$, or $\beta_k^{(n)}(a)$, respectively, for arbitrary a (often, however, we shall have to restrict ourselves to values $a \geq 2$). From these, of course, analogous results for the zeros of $y_n(z;a,b)$, or $\theta_n(z;a,b)$ (which we may denote by $\alpha_k^{(n)}(a,b)$ and $\beta_k^{(n)}(a,b)$, respectively) can be obtained immediately by a change of variable and we shall usually refrain from writing out explicit results for $b \neq 2$. Also, whenever possible the subscripts and/or superscripts will be suppressed.

From

$$(1) \qquad z^n y_n(1/z) = \theta_n(z) = e^z \phi_n(z)$$

it is clear that the zeros of $\theta_n(z)$ and of $\phi_n(z)$ are the same and are the reciprocals of the zeros of $y_n(z)$, so that $\alpha_k \beta_k = 1$ $(k = 1,2,\ldots,n)$. These considerations remain valid even for $a \neq 2 \neq b$, in fact, even for $\phi_n(z;a,b,c)$. Hence, any information about the zeros of any one of these functions can be translated immediately into information about the zeros of all of them.

Some of the properties of the zeros of BP have attracted the attention quite early (see [17] and [53]). Others, such as good upper and lower bounds for $|\alpha_k|$, or $|\beta_k|$, respectively, were discovered and improved only later.

2. Some of the oldest known results are formulated in the following three theorems.

<u>THEOREM 1</u>. ([17], [53], see also [47].) (a) *All zeros* α_k $(k = 1,2,\ldots,n)$ *of* $y_n(z)$ *(or* β_k *of* $\theta_n(z)$*) are simple.* (b). *No two consecutive polynomials* $y_n(z)$, $y_{n+1}(z)$ *(or* $\theta_n(z)$, $\theta_{n+1}(z)$, *respectively) have a zero in common.* (c) *No* BP *of even degree has a real zero and those of odd degree have exactly one real, negative zero.*

<u>THEOREM 2</u>. (see [53]). (a) *All zeros of the* BP $y_n(z)$ *satisfy* $|\alpha_k^{(n)}| < 1$, *except for* $\alpha_1^{(1)} = -1$; *correspondingly,* $|\beta_k^{(n)}| > 1$, *except for* $\beta_1^{(1)} = -1$. (b) *The real zeros* $\alpha_m^{(2m-1)}$ *of the polynomials* $y_{2m-1}(z)$ *form a monotonically increasing sequence:*

$-1 = \alpha_1^{(1)} < \alpha_2^{(3)} < \dots < \alpha_m^{(2m-1)} < 0.$

<u>THEOREM 3</u>. ([17]; see also [62] and [2]). *The zeros* $\alpha_r = \alpha_r^{(n)}$ *of* $y_n(z)$ *satisfy the equations*

(2) $$\sum_{r=1}^{n} \alpha_r = -1, \ \sum_{r=1}^{n} \alpha_r^{2m-1} = 0 \text{ for } m = 2,3,\dots,n.$$

Correspondingly, the zeros $\beta_r = \beta_r^{(n)}$ *of* $\theta_n(z)$ *satisfy*

(2') $$\sum_{r=1}^{n} \beta_r^{-1} = -1, \ \sum_{r=1}^{n} \beta_r^{1-2m} = 0 \text{ for } m = 2,3,\dots,n.$$

3. <u>Proof of Theorem 1</u>. (a) If α is a multiple zero of $y_n(z)$, then $y_n(\alpha) = y_n'(\alpha) = 0$. By (3.10), also $y_{n-1}(\alpha) = 0$ and by (3.16) it now follows that $y_{n-1}'(\alpha) = 0$ Hence, if α is a multiple zero of $y_n(z)$, it is also a multiple zero of $y_{n-1}(z)$ and so, by induction on n, a multiple zero of $y_1(z) = z+1$, which is absurd. (b) By (3.7) it follows that if α is a common zero of $y_{n+1}(z)$ and $y_n(z)$, then it is also a zero of $y_{n-1}(z)$. By induction on n it is also a zero of $y_2(z)$ and of $y_1(z)$, i.e. $\alpha = -1$. However, $y_2(-1) \neq 0$ and the assertion follows. (c) This assertion follows almost immediately from the theory of the Bessel functions $K_\nu(z)$ for $\nu = n + 1/2$ (see, e.g. [1], p. 377 and Fig. 96), but can also be proved directly, as follows. By (1) it is sufficient to prove the statement for $\phi_n(z)$. We recall that $f(z) = \phi_n(z)$ and $g(z) = \phi_n(-z)$ are both solutions (easily seen to be independent) of (2.13). It follows that

$$zf'' - 2nf' = zf,$$

$$zg'' - 2ng' = zg,$$

whence $z\frac{d}{dz}(f'g - fg') = z(gf'' - fg'') = 2n(f'g - fg')$, or, by integration,

(3) $$f'g - fg' = Cz^{2n}.$$

While irrelevant for our present purpose,we observe that (2.2b) and (2.5) lead to $\lim_{z\to\infty} z^{-2n}(f'g-fg') = 2(-1)^{n+1}$ and (3) shows that this value remains unchanged for all z, so that $C = 2(-1)^{n+1}$.

We now complete the proof of (c), by following Burchnall [17]. We return to $\theta_n(z)$, replace f and g in (3) by their respective values and obtain

$\frac{d}{dz}(e^{-z}\theta_n(z))\cdot(e^{z}\theta_n(-z))-(e^{-z}\theta_n(z))\cdot\frac{d}{dz}(e^{z}\theta_n(-z)) = (-1)^{n+1}\cdot 2z^{2n}$. This simplifies to (see [17])

$$(4) \qquad -2\theta_n(z)\theta_n(-z)+\theta_n'(z)\theta_n(-z)+\theta_n(z)\theta'(-z) = 2(-1)^{n+1}z^{2n}.$$

The coefficients of $\theta_n(z)$ are all positive. Hence, any real zero of $\theta_n(z)$ is negative Let $-\alpha$, $-\beta(\alpha > 0,\ \beta > 0)$ be two consecutive real zeros (assuming that such exist). Then (4) shows that

$$\theta_n'(-\alpha)\theta_n(\alpha) = 2(-1)^{n+1}\alpha^{2n},$$

and similarly

$$\theta_n'(-\beta)\theta_n(\beta) = 2(-1)^{n+1}\beta^{2n}.$$

As $\theta_n(\alpha)$ and $\theta_n(\beta)$ are both positive, $\theta_n'(-\alpha)$ and $\theta_n'(-\beta)$ have the same sign, which is impossible for consecutive zeros. Hence, the assumption that there exist two real zeros is false. Finally, the reality of the coefficients of $\theta_n(z)$ shows that complex zeros occur as complex, conjugate pairs and Theorem 1 is completely proved.

4. Proof of Theorem 2. The proof of Theorem 2 is based on a Theorem of Eneström [14] and Kakeya [40]. We shall state it in the stronger version due to Hurwitz [34], in the form of a sequence of three Theorems. For completeness, we shall sketch their proofs, but the reader, willing to accept their statement, may skip these.

THEOREM A. *Let* $b_0 \geq b_1 \geq \dots \geq b_n > 0$; *then the polynomial* $f(z) = \sum_{j=0}^{n} b_j z^j$ *has no zero with* $|z| < 1$.

Proof. Let $|z| = r < 1$; then, with $b_{-1} = b_{n+1} = 0$, $|f(z)(1-z)| =$

$$\left|b_o - \sum_{j=1}^{n} (b_{j-1}-b_j)z^j - b_n z^{n+1}\right| \geq b_0 - \sum_{j=1}^{n} (b_{j-1}-b_j)r^j - b_n r^{n+1} = -\sum_{j=0}^{n+1} (b_{j-1}-b_j)r^j;$$

consequently, $|f(z)(1-z)| \geq -\sum_{j=0}^{n-1} (b_{j-1}-b_j)r^j = f(r)\cdot(1-r) > 0$.

THEOREM B. (Kakeya). *Let the coefficients* b_j *of* $f(z)$ *be positive and set* $\operatorname{Max}_j (b_j/b_{j+1}) = \rho$, $\min_j (b_j/b_{j+1}) = \sigma$; *then all zeros* z_k $(k = 1,2,\dots,n)$ *of* $f(z)$ *satisfy the inequalities* $\sigma \leq |z_k| \leq \rho$.

Proof. Consider $f(z\rho) = \sum_{j=0}^{n} b_j\rho^j z^j = g(z)$. Then $\dfrac{b_j\rho^j}{b_{j+1}\rho^{j+1}} \leq \dfrac{1}{\rho} \operatorname{Max}_j \dfrac{b_j}{b_{j+1}} = 1$; hence, the coefficients of $g(z)$ are increasing, those of $h(z) = z^n g(1/z)$ are decreasing and Theorem A applies to $h(z)$. If u is a zero of $h(z)$, $1/u$ is a zero of $g(z)$ and ρ/u is a zero of $f(z)$. By Theorem A, $|u| \geq 1$, so that any zero z_k of $f(z)$ satisfies

$|z_k| = |\rho/u| \leq \rho$, as claimed. The other inequality follows in the same way by consideration of $z^n f(\sigma/z)$.

THEOREM C. (Hurwitz). *The equation* $f(z) = 0$ *with positive, monotonically decreasing coefficients has a zero of absolute value equal to one, if and only if the coefficients can be partitioned into sets of* m *consecutive equal coefficients, where* m *is a divisor of* n+1.

Proof. For a zero z, from $f(z)(1-z) = -\sum_{j=0}^{n+1} (b_{j-1}-b_j)z^j = 0$ follows $\sum_{j=1}^{n+1} (b_{j-1}-b_j)z^j = b_o$. For $|z| = 1$ also follows

$$\sum_{j=1}^{n+1} |(b_{j-1}-b_j)z^j| = \sum_{j=1}^{n+1} (b_{j-1}-b_j) = b_o = |\sum_{j=1}^{n+1} (b_{j-1}-b_j)z^j|.$$

Consequently, all summands have the same argument, i.e. $(b_{j-1}-b_j)z^j = p_j e^{i\alpha}$, α real, $p_j \geq 0$. In fact, $\alpha = 0$, because $\sum_{j=1}^{n+1} (b_{j-1}-b_j)z^j = e^{i\alpha} \sum_{j=1}^{n+1} p_j = b_o$. From $b_n z^{n+1} = p_n > 0$ now follows $z^{n+1} > 0$ and using also $|z| = 1$, $z^{n+1} = 1$. Let m be the smallest integer such that $z^m = 1$; then m divides n+1 and $m \geq 2$ (otherwise $z = 1$ and $0 = \sum_{j=0}^{n} b_j z^j = \sum_{j=0}^{n} b_j > 0$, a contradiction). From $(b_{j-1}-b_j)z^j = p_j \geq 0$ follows, for $m\nmid j$, that $b_{j-1}-b_j = 0$. In particular, $b_o = b_1 = \ldots = b_{m-1}$; $b_m = b_{m+1} = \ldots = b_{2m-1}$, etc. which proves the necessity of the condition. On the other hand, if the condition on the b_j's holds, then $f(z) = (\sum_{k=0}^{t} b_{km}z^{km})(\sum_{j=0}^{m-1} z^j)$, with $t = (n+1-m)/m$, so that $f(e^{2\pi ir/m}) = 0$ for $r = 1,2,\ldots,m-1$, and this proves the sufficiency. The proof of Theorem 2(a) is now immediate. From (2.8) it follows that the coefficients of $\theta_n(z) = \sum_{m=0}^{n} a_{n-m}z^m$ are decreasing from $a_n = (2n)!/2^n n!$ to $a_o = 1$. Hence, Theorem A is applicable with $b_m = a_{n-m}$ and all zeros β_k of $\theta_n(z)$ satisfy $|\beta_k| \geq 1$. Going beyond this, we verify that $r_m = \frac{a_{n-m}}{a_{n-m-1}} = \frac{b_m}{b_{m+1}} = \frac{1}{2}\ \frac{(m+1)(2n-m)}{n-m}$ $(m = 0,1,\ldots,n-1)$, with $\rho = \operatorname{Max}_m r_m = n(n+1)/2$ and $\sigma = \min_m r_m = 1$.

Theorem 2(a) is now a simple corollary of the stronger

THEOREM 4. *For* $n > 1$, *all zeros* $\beta_k = \beta_k^{(n)}$ *of* $\theta_n(z)$ *satisfy* $1 < |\beta_k| \le n(n+1)/2$. *The zeros* α_k *of* $y_n(z)$ $(n > 1)$ *satisfy* $2/n(n+1) \le |\alpha_k| < 1$. *For* $n = 1$, *the unique zero of* $\theta_1(z)$ *and of* $y_1(z)$ *is* $\alpha_1^{(1)} = \beta_1^{(1)} = -1$.

Proof. On account of $\rho = n(n+1)/2$, $\sigma = 1$ and of Theorem B, the only statements of Theorem 4 still to be justified are the omissions of equal signs for $n > 1$. For that we use Theorem C. The equal sign requires that all coefficients fall into sets of at least two consecutive equal coefficients. However, the ratio $r_m = \frac{a_{n-m}}{a_{n-m-1}}$ equals $(m+1)(2n-m)/(2n-2m) \ge m+1 \ge 1$, with equality only for $m = 0$, when $r_0 = 1$. Hence, equality can exist only between the first two coefficients. The sets of consecutive equal coefficients required by Theorem C reduce, consequently, to a single set of two coefficients, so that $n = 1$, when $y_n(z) = \theta_1(z) = z+1$, and when we have indeed $\alpha_1 = \beta_1 = -1$.

Theorem 4 and with it Theorem 2(a) is proved.

The proof of Theorem 2(b) is based on (3.7). Let $\alpha^{(n)}$ be the unique real zero of $y_n(z)$ (n odd). Then (3.7) with n-1 instead of n and $z = \alpha^{(n-2)}$ becomes $y_n(\alpha^{(n-2)}) = (2n-1)\alpha^{(n-2)}y_{n-1}(\alpha^{(n-2)})$. By Theorem 1, $\alpha^{(n-2)} < 0$ and $y_{n-1}(\alpha^{(n-2)}) > 0$ because n-1 is even, $y_{n-1}(z)$ does not change its sign and $y_{n-1}(0) = 1 > 0$; consequently, $y_n(\alpha^{(n-2)}) < 0$, $y_n(\alpha^{(n)}) = 0$ and, by Theorem 1, $y_n(z) > 0$ for $z > \alpha^{(n)}$, so that $\alpha^{(n-2)} < \alpha^{(n)} < 0$, as claimed.

5. Proof of Theorem 3. (see [17]). By logarithmic differentiation of (1) we obtain, successively:

$$\frac{\phi_n'(z)}{\phi_n(z)} + 1 = \frac{\theta_n'(z)}{\theta_n(z)} = \sum_{r=1}^{n} \frac{1}{z-\beta_r} = -\sum_{r=1}^{n} \frac{1}{\beta_r}\,\frac{1}{1-z/\beta_r} = -\sum_{r=1}^{n}\sum_{k=1}^{\infty} \frac{z^{k-1}}{\beta_r^k}$$

$$(5)\qquad = -\sum_{k=1}^{\infty} z^{k-1}\sigma_{-k}, \text{ where } \sigma_m = \sum_{r=1}^{n} \beta_r^m.$$

By Theorem (2.3), $\phi_n(z)$ contains no odd powers of z with exponent less than 2n+1; it follows that neither $\phi_n'(z)$, nor $\phi_n'(z)/\phi_n(z)$ contains even powers of exponent less than 2n. Hence, by comparing coefficients in (5) for k odd, $0 < k-1 < 2n$, one has $\sigma_{-k} = 0$. For $k = 1$, by (5), $1 = -\sigma_{-1}$. Consequently, $\sum_{r=1}^{n} \beta_r^{1-2m} = 0$ for $m = 2,3,\dots,n$, while $\sum_{r=1}^{n} \beta_r^{-1} = -1$. We recall also that the α_r's are the reciprocals of the β_r's and with this Theorem 3 is proved. For a proof of the converse, namely that every solution of the system (2) (or (2')) consists of the zeros $\beta_r^{(n)}$ of $\theta_n(z)$ (or $\alpha_r^{(n)}$ of $y_n(z)$, respectively) taken in some order, see [17].

6. It is easy to improve some of previous results. First we observe that the proofs of Theorem 1(a) and (b) go through for $y_n(z;a)$, provided only that we use (3.21) instead of (3.10); (3.22) instead of (3.16).

Next, the statement of Theorem 1(b) can be strengthened to read as follows (see [47]).

THEOREM 1(b'). *If* $n \neq m$, *then no zero of* $y_n(z)$ *can be a zero of* $y_m(z)$; *in fact, no zero of* $y_n(z;a)$ *can be a zero of* $y_m(z;a)$.

Proof. For the proof of the first statement (see [47]), it is sufficient to use (3.7') instead of (3.7) and for the second one to use the obvious generalization of (3.20) instead of (3.7').

We have proved

THEOREM 1'. *The statements of Theorem* 1(a) *and* 1(b') *hold for the generalized* BP.

REMARK. The proof of Theorem 1(c) does not translate immediately into one for $y_n(z;a)$. The problem of determining the values of a for which Theorem 1(c) holds appears to be open.

Also Theorem 2(a) can be extended to the case $a \neq 2$. Indeed, by (2.27), with the coefficients $d_k = d_k^{(n)}$ of $\theta_n(z;a)$, $r_m = \frac{d_{n-m}}{d_{n-m-1}} = \frac{b_m}{b_{m+1}} = \frac{(m+1)(2n-m+a-2)}{2(n-m)}$. If we consider for a moment m as a continuous variable and take the derivative, we verify that in general r_m increases monotonically with m, from $\sigma(a) = 1 + \frac{a-2}{2n}$ for $m = 0$ to $\rho(a) = n(n+a-1)/2$ for $m = n-1$. An exception can occur only for $n < -a+2$, i.e., only for $a < 2$, and even in those cases only for finitely many values of n. As these can be studied individually, we shall ignore them here and assume that $n \geq -a+2$. By using Theorems A and B as before we obtain

THEOREM 2'. *For* $n \geq \mathrm{Max}(-a+2,2)$ (*i.e., if* $a \geq 0$, *for all* $n \geq 2$)

$$1 + (a-2)/2n \leq |\beta_k(a)| \leq n(n+a-1)/2,$$

$$2/n(n+a-1) \leq |\alpha_k(a)| \leq 2n/(2n+a-2).$$

For $n = 1$, *the unique zero of* $y_1(z;a)$ *is* $\alpha_1^{(1)} = -2/a$ *and the zero of* $\theta_1(z;a)$ *is* $\beta_1^{(1)} = -a/2$.

7. In previously mentioned paper [47], Dickinson, by using the connection of BP with Lommel polynomials (see Chapter 5), showed that no $\alpha_k^{(n)}$ is purely imaginary. We shall use a different method (see [12]) to prove the stronger result (see [111] and [95]) of

THEOREM 5. *For all* n, *the zeros* $\alpha_k^{(n)}$ *of* $y_n(z)$ *satisfy* $\mathrm{Re}\ \alpha_k^{(n)} < 0$.

In the proof we use the following known theorem.

THEOREM D. *Let* $P(z) = \sum_{j=0}^{n} a_j z^{n-j}$ *and set* $Q(z) = \sum_{k=1}^{(n+1)/2} a_{2k-1} z^{n-2k+1}$. *All the zeros of* P(z) *have negative real parts, if and only if, in the notation of Chapter* 8, $Q(z)/P(z) = [0, c_1z+1, c_2z, \ldots, c_nz]$ *with the coefficients* c_j $(j = 1,2,\ldots,n)$ *all positive.*

For a proof of Theorem D, see, e.g. [111].

Proof of Theorem 5. In the notations of Chapter 8,

$$y_n(z) = P_n(z) + Q_n(z), \text{ with } Q_n = \tfrac{1}{2}\{y_n(z) - (-1)^n y_n(-z)\}.$$

By Theorem 8.1, $\dfrac{y_n(z)}{Q_n(z)} = 1 + \dfrac{P_n(z)}{Q_n(z)}$. By Theorem 8.2 this is the n-th convergent of

$$1 + \frac{e^{2/z}+1}{e^{2/z}-1} = 1 + [z, 3z, \ldots, (2n-1)z, \ldots] = [1+z, 3z, \ldots, (2n-1)z, \ldots].$$

Hence,

$$\frac{Q_n(z)}{y_n(z)} = [0, 1+z, 3z, \ldots, (2n-1)z].$$

As already observed in Chapter 8, $Q_n(z)$ contains precisely all powers of $y_n(z)$ of parity opposite to n, so that we may identify $Q_n(z)$ and $y_n(z)$, with Q(z) and P(z) respectively of Theorem D and the result follows.

Wimp [112] proved that Theorem 5 holds also for the zeros $\alpha_k^{(n)}(a)$ of $y_n(z;a)$ at least for $a \geq 2$.

8. Another result of Dickinson [47], namely that the origin is a limit point of zeros of BP $y_n(z)$, as well as Nasif's [81] bound $|\alpha_k^{(n)}| \leq \{(n-1)/(2n-1)\}^{1/2}$ and McCarthy's [74] generalizations of some of these results are all consequences of a result of Dočev [48], that we shall prove directly in the slightly more general form of

THEOREM 6. *For any complex* a *and positive* b *the zeros* $\alpha_k^{(n)}(a,b)$ *of* $y_n(z;a,b)$ *satisfy the inequality*

$$|\alpha_k^{(n)}(a,b)| \leq \frac{b}{n-1+\mathrm{Re}\ a}.$$

COROLLARY 1. *For* $a = b = 2$, $|\alpha_k^{(n)}| \leq 2/(n+1)$.

We observe that this result is sharp, because for $n = 1$ it holds with the equal sign. The upper bound of Theorem 6 may be combined, with the lower bound of Theorem 2' and the restriction of Theorem 5 in order to lead us close to what are essentially the best presently known statements concerning the locations of the zeros of $y_n(z;a)$ and $\theta_n(z;a)$, respectively, that are valid for all n. See, however, besides [85] also [98], [46], and [84], the latter with better results, but valid, only asymptotically, for $n \to \infty$.

THEOREM 7. *For* $a \geq 2$, *all zeros* $\alpha_k(a)$ *of* $y_n(z;a)$ *belong to the semi-annulus defined by the inequalities:*

$$\frac{2}{n(n+a-1)} \leq |\alpha_k(a)| \leq \frac{2}{n+a-1}, \quad \mathrm{Re}\ \alpha_k < 0.$$

The zeros $\beta_k(a)$ *of* $\theta_n(z;a)$ *similarly satisfy*

$$(n+a-1)/2 \leq |\beta_k(a)| \leq n(n+a-1)/2, \quad \mathrm{Re}\ \beta_k < 0.$$

The inequalities for $|\alpha_k(a)|$ *and* $|\beta_k(a)|$ *are sharp and become equalities for* $n = 1$.

COROLLARY 2. *The zeros* α_k *of* $y_n(z)$ *and the zeros* β_k *of* $\theta_n(z)$, *belong to the semi-annuli*

$$\frac{2}{n(n+1)} \leq |\alpha_k| \leq \frac{2}{n+1}, \quad \mathrm{Re}\ \alpha_k < 0$$

and

$$(n+1)/2 \leq |\beta_k| \leq n(n+1)/2, \quad \mathrm{Re}\ \beta_k < 0,$$

respectively.

Theorem 7 and its corollary are immediate consequences of previously quoted results, of which only Theorem 6 remains to be proved. The proof relies heavily on Theorem E, due to Laguerre. Up to now most theorems not directly connected with BP and whose proof requires more than just a few lines have only been stated for ease of reference and with an indication where a proof may be found. It is, however, not entirely trivial to formulate Laguerre's theorem (first published in 1882; see [41]) in the form here needed. In fact, the clearest presentation that the author could find is in a sequence of problems in Pólya and Szegö's "Aufgaben und Lehrsätze aus der Analyse" ([49], vol. 2, Problems 102-118). For these reasons the proof of Laguerre's theorem will be sketched here.

9. THEOREM E. (Laguerre [**41**]). *Let* f(x) *be a polynomial of degree* n *and set* $X(x) = x-2(n-1)f'(x)/f''(x)$. *Let* x_0 *be a simple zero of* f(x) *and consider a circle* C *(possible a straight line) through the point represented by the complex number* x_0*, such that no zero of* f(x) *belongs to one of the two open, circular regions determined by* C. *Then* $X(x_0)$ *belongs to the other region.*

Proof of Theorem E. Let $z_1, z_2, \ldots, z_n$ be complex numbers with the center of mass $\zeta = (\sum_{j=1}^{n} z_j)/n$. We generalize this concept and shall say that $\zeta = \zeta_\infty$ as just defined is the center of mass of the points z_j $(j = 1,\ldots,n)$ with respect to the "pole" $z_0 = \infty$. In order to define the center of mass ζ_{z_0} of the z_j's with respect to an arbitrary pole z_0, different from all the points z_j, we proceed as follows: First, we map z_0 into the point at infinity, by a linear fractional transformation, say $z' = \frac{a}{z-z_0} + b$ (a,b arbitrary complex numbers). Under this map z_j is sent into $z'_j = \frac{a}{z_j-z_0} + b$. We now find $\zeta' = \zeta'_\infty = (\sum_{j=1}^{n} z'_j)/n$ as before and then map ζ' back. The inverse of ζ' under our mapping will be denoted by $\zeta = \zeta_{z_0}$ and is the desired generalized center of mass of the z_j's with respect to z_0. We now show that ζ_{z_0} depends only on z_0 and the z_j's and is independent of the specific auxiliary map selected. Indeed, $\sum_j z'_j = a\sum_j (z_j-z_0)^{-1} + nb$, so that $\zeta'_\infty = \frac{a}{n}\sum_j \frac{1}{z_j-z_0} + b$. However, also $\zeta'_\infty = \frac{a}{\zeta_{z_0}-z_0} + b$ holds, so that $\frac{1}{n}\sum_j \frac{1}{z_j-z_0} = \frac{1}{\zeta_{z_0}-z_0}$, and

$$\zeta_{z_0} = z_0 + n/\sum_j (z_j-z_0)^{-1} = z_0 - n/\sum_j (z_0-z_j)^{-1}$$

is indeed independent of a and b.

Next, let $F(z) = \prod_{j=1}^{n} (z-z_j)$ be a polynomial. Then $\frac{F'}{F}(z) = \sum_j \frac{1}{z-z_j}$, so that the center ζ_z of the zeros with respect to an arbitrary pole z is given by

$$\zeta_z = z - nF(z)/F'(z). \tag{6}$$

From the Gauss-Lucas theorem we know that the (ordinary) center of mass ζ_∞ of the zeros of F(z) is inside the convex closure of those zeros and that this is the smallest convex set containing all z_i's. Each side of the polygon is a straight line through two zeros. It may be considered also as a circle through any two zeros and the pole $z = \infty$. It follows that ζ'_∞ is inside the corresponding polygon of the (z_j')'s. When we map back, the sides of that polygon become circles through any two of the z_j's and the pole z. Of the two circular domains determined by such a circle, it may happen that one contains no other zeros of $F(z)$. Let us "delete" it then mentally and keep only the one that contains all other zeros. The intersection of all these remaining circular regions (the image of the intersections of the corresponding half planes through the (z_j')'s, taken two by two) yields a curvilinear polygon, say C_z, that contains ζ_z, the image of ζ'_∞.

Claim 1. If the set of zeros z_j $(j = 1,2,\ldots,n)$ belongs to any circular domain D and if z is outside D, then C_z is in D.

Proof of Claim 1. The smallest convex polygon containing all the (z_j')'s is inside any circle D', containing all the (z_j')'s (because that polygon is the smallest convex set containing all (z_j')'s and a circle is a convex set), while $z' = \infty$ is outside D'. When we map back, set inclusions are preserved and Claim 1 is proved.

Claim 2. Let x be a simple zero of $F(z)$. The center of mass of the remaining zeros with respect to x is

$$X = x-2(n-1)F'(x)/F''(x). \tag{7}$$

Proof of Claim 2. Let $F(z) = (z-x)f(z)$; then $F'(z) = f(z) + (z-x)f'(z)$, $F''(z) = 2f'(z) + (z-x)f''(z)$, so that $F'(z) = f(x)$ and $F''(x) = 2f(x)$.

We now apply (6) to $f(z)$ (whose degree is $n-1$ with respect to x) and obtain $X = x-(n-1)f(x)/f'(x)$. If we replace $f(x)$ and $f'(x)$ by $F'(x)$ and $F''(x)/2$, respectively, we obtain (7) and Claim 2 is proved.

Let D_1 be a circle through x that contains all other zeros of $F(z)$. As x is a simple zero, one can deform D_1 into a circle D that leaves x outside, but still contains all other zeros. Let C_x be the curvilinear polygon of these other zeros of $F(z)$, with respect to x. We know from Claim 1 that $X = X(x)$ belongs to C_x and $C_x \subset D$ because $x \notin D$. This finishes the proof of Laguerre's Theorem E.

10. From Theorem E immediately follows the

COROLLARY 3. *If* x_0 *is one of the zeros of largest modulus of a polynomial* $f(x)$, *then* $|X(x_0)| \le |x_0|$.

Proof. No zero belongs to the open circular domain $|z| > |x_0|$; hence, $|X(x_0)| > |x_0|$ is not possible and the conclusion follows.

LEMMA 1. *If the polynomial* f(z) *satisfies a linear differential equation of second order of the form*

$$P(z)y'' + Q(z)y' + R(z)y = 0,$$

then, for each zero z_0 *of* f(z),

$$\frac{f'(z_0)}{f''(z_0)} = -\frac{P(z_0)}{Q(z_0)}. \tag{8}$$

Proof. This follows trivially from $f(z_0) = y(z_0) = 0$.

LEMMA 2. *Let* α *be a zero of* $y_n(z;a,b)$; *then*

$$X(\alpha) = \alpha + \frac{2(n-1)\alpha^2}{a\alpha + b}.$$

Proof. By (2.26), $y_n(z;a,b)$ satisfies a linear differential equation as described in Lemma 1, with $P(z) = z^2$ and $Q(z) = az+b$, so that the result follows from (7) and (8).

LEMMA 3. *If* α *is one of the zeros of largest modulus of* $y_n(z;a,b)$ *then*

$$\left|1 + \frac{2(n-1)}{a+b/\alpha}\right| \le 1.$$

Proof. By Corollary 3, if α is a zero of maximum modulus, then $|X(\alpha)| \le |\alpha|$ and the result follows by Lemma 2.

Proof of Theorem 6. Let $v = \alpha^{-1}$, where α is one of the zeros of largest modulus of $y_n(z;a,b)$. Then, by Lemma 3, $|1+2(n-1)/(a+bv)| \le 1$. We start by determining the locus of the complex variable v, for which we obtain equality. It is given by

$$1 + \frac{2(n-1)}{a+bv} = e^{i\phi},\ 0 \le \phi < 2\pi, \tag{9}$$

or, solving for v, by

$$\begin{aligned} v &= -b^{-1}(a+n-1+i(n-1)\operatorname{cotg}\phi/2) \\ &= -b^{-1}\{\operatorname{Re} a+n-1+i\ ((n-1)\operatorname{cotg}\phi/2 + \operatorname{Im} a)\}. \end{aligned}$$

As ϕ varies from 0 to 2π, v describes the straight line L parallel to the imaginary axis, of abscissa $-(n+\operatorname{Re} a-1)/b$. Consequently, the locus of v^{-1} for which one obtains the equality (9), is the inversion of L in the unit circle. This is the circle through the origin, with center on the real axis and passing through $-b/(n+\operatorname{Re} a-1)$. Its equation is

$$\left|z + \frac{b}{2(n+\mathrm{Re}\ a-1)}\right| = \frac{b}{2(n+\mathrm{Re}\ a-1)} .$$

The required inequality for α corresponds to points inside the circle. In particular,

$$|\alpha| \leq b/(n+\mathrm{Re}\ a-1)$$

and Theorem 6 (hence, also Theorem 7) is proved.

REMARK. Dočev considers only the case $b = -1$ and denotes $a-2$ by m.

11. In 1976 appeared a paper by Saff and Varga [98] with an important improvement of Dočev's result. It became available too late for a complete treatment here; therefore, only a brief outline will be given, although it leads to the strongest presently known results concerning the location of the zeros of $y_n(z;a)$ and $\theta_n(z;a)$.

The method is based on the following

THEOREM F. *Let* $\{p_n(z)\}_{n=0}^{k}$ *be a sequence of polynomials of respective degrees* n, *which satisfy the three-term recurrence relation*

$$(10) \qquad p_n(z) = \left(\frac{z}{b_n} + 1\right)p_{n-1}(z) - \frac{z}{c_n}\, p_{n-2}(z) \qquad (n = 1,2,\ldots,k),$$

where the b_n*'s and* c_n*'s are positive real numbers and where* $p_{-1}(z) = 0$, $p_0(z) = p_0 \neq 0$. *Set* $\alpha = \min_{1 \leq n \leq k} b_n(1-b_{n-1}c_n^{-1})$ $(b_0 = 0)$. *If* $\alpha > 0$, *then the parabolic region*

$$P_\alpha = \{z = x+iy \in \mathbf{C} \,|\, y^2 \leq 4\alpha(x+\alpha),\ x > -\alpha\}$$

contains no zeros of any of the polynomials $p_n(z)$, $n = 1,2,\ldots,k$.

Sketch of Proof. Let $z_0 \in P_\alpha$, such that none of the $p_n(z)$ vanishes at z_0 and set $\mu_n(z) = zp_{n-1}(z)/b_np_n(z)$ $(n = 1,2,\ldots,k)$. First we verify that $\mu_n = \mu_n(z_0)$ has $\mathrm{Re}\ \mu_n \leq 1$ $(n = 1,2,\ldots,k)$. This is clear for $\mu_1 = z_0/(z_0+b_1)$, because in P_α, $\mathrm{Re}\ z \geq -b_1$; and for $1 < n \leq k$ that property follows by induction on k, by use of (10) and the definition of P_α. Next, we observe that $p_n(0) \neq 0$ for all $n = 1,2,\ldots,k$. Indeed, by (10), $p_n(0) = p_{n-1}(0)$; hence, by induction, for every n, $p_n(0) = p_0(0) = p_0 \neq 0$. It follows that if $p_n(z_0) = 0$, then $z_0 \neq 0$. Assume now that $p_n(z_0) = p_{n-1}(z_0) = 0$; then, by (10) and $z_0 \neq 0$, also $p_{n-2}(z_0) = 0$ and, by induction $p_{n-j}(z_0) = 0$ for all $j \leq n$, and this if false for $j = n$. Consequently, $p_n(z)$ and $p_{n-1}(z)$ have no common zeros. Let now, contrary to the statement of Theorem F, $p_n(z_0) = 0$ for some n in $1 \leq n \leq k$ and $z_0 \in P_\alpha$. By (10), $p_1(z) = (z+b_1)p_0/b_1$, so that $z = -b_1$ is the (only) zero of $p_1(z)$. By the definition of α, $-b_1 \leq -\alpha$, so that $-b \notin P_\alpha$; hence, $2 \leq n \leq k$. Again by (10),

$$(z_0 b_n^{-1}+1)p_{n-1}(z_0) = z_0 c_n^{-1} p_{n-2}(z_0), \text{ or } \frac{z_0+b_n}{b_n} = \frac{z_0}{c_n}\frac{p_{n-2}(z_0)}{p_{n-1}(z_0)};$$

this is meaningful, because $p_n(z_0) = 0$ implies, as seen, that $p_{n-1}(z_0) \neq 0$. This shows that $\frac{c_n}{b_{n-1}b_n}(z_0+b_n) = \frac{z_0 p_{n-2}(z_0)}{b_{n-1}p_{n-1}(z_0)} = \mu_{n-1}(z_0)$ and, by $z_0 \in P_\alpha$, Re $\mu_{n-1}(z_0) \leq 1$.

Hence, Re $\frac{c_n}{b_{n-1}b_n}(z_0+b_n) \leq 1$, or $\frac{c_n}{b_{n-1}b_n}$ Re $z_0 \leq 1 - \frac{c_n}{b_{n-1}}$, whence

Re $z_0 \leq -b_n(1-b_{n-1}c_n^{-1}) \leq -\alpha$. This, however, shows that z_0 could not have been in P_α, where for all z, Re $z > -\alpha$. Theorem F is proved.

Consider now the polynomials $\theta_n(z;a) = \sum_{m=0}^{n} d_m^{(n)} z^{n-m}$, with $d_m^{(n)} = \frac{n!(n+m+a-2)^{(m)}}{m!(n-m)!2^m}$, as in (2.27). By comparing the coefficients of z^{n-m} $(m = 0,1,\ldots,n)$ on both sides, one verifies that

$$\theta_n(z;a) = (z+n-1+a/2)\theta_{n-1}(z;a+1) - \tfrac{1}{2}(n-1)z\,\theta_{n-2}(z;a+2). \tag{11}$$

For each polynomial in (11), the sum of the degree and the a-entry is the same, say $n+a = s$. With $a = s-n$, (11) becomes

$$\theta_n(z;s-n) = (z+(s+n-2)/2)\theta_{n-1}(z;s-(n-1)) - \tfrac{1}{2}(n-1)z\,\theta_{n-2}(z;s-(n-2)).$$

If we set $b_n = (s+n-2)/2$ and divide by b_n, we obtain

$$2(s+n-2)^{-1}\theta_n(z;s-n) = (zb_n^{-1}+1)\theta_{n-1}(z;s-(n-1))-(n-1)(s+n-2)^{-1}z\theta_{n-2}(z;s-(n-2)). \tag{12}$$

Equation (12) is almost of the required form (10), except for the factor $2/(s+n-2)$ of $\theta_n(z;s-n)$. To eliminate this discrepancy, we multiply (12) by $2^{n-1}/(s+n-3)^{(n-2)}$ and obtain

$$\frac{2^n}{(s+n-2)(s+n-3)\ldots s}\theta_n(z,s-n) =$$

$$(\frac{z}{b_n}+1)\frac{2^{n-1}}{(s+n-3)\ldots s}\theta_{n-1}(z;s-(n-1)) - \frac{z}{c_n}\frac{2^{n-2}}{(s+n-4)\ldots s}\theta_{n-2}(z;s-(n-2)),$$

where we have set $c_n = (s+n-2)(s+n-3)/2(n-1)$. This is precisely of the form (10), with $p_n(z) = \frac{2^n}{(s+n-2)^{(n-1)}}\theta_n(z;s-n)$ and b_n, c_n as defined. Also,

$b_n(1-b_{n-1}c_n^{-1}) = (s-1)/2$ independently of n, so that $\alpha = (s-1)/2$. By Theorem F, none of the polynomials $p_n(z)$, hence none of the polynomials $\theta_n(z;s-n)$, can have zeros inside the parabola $y^2 \le 2(s-1)(x+(s-1)/2) = (s-1)(2x+s-1)$. This may be written as $|z|^2 = x^2+y^2 \le x^2+2(s-1)x+(s-1)^2 = (x+s-1)^2$, or $|z| \le \mathrm{Re}\, z + s-1$. In polar coordinates $r \le r\cos\theta+s-1$, or $r \le \frac{s-1}{1-\cos\theta} = \frac{n+a-1}{1-\cos\theta}$. Consequently, for all zeros $\beta_k^{(n)}(a)$ $(k = 1,2,\dots,n-1)$ of $\theta_n(z;a)$ we have $|\beta_k^{(n)}(a)| > \frac{n+a-1}{1-\cos\theta}$, and therefore the zeros $\alpha_k^{(n)}(a)$ of $y_n(z;a)$ are inside the cardioid $|\alpha_k^{(n)}(a)| < (1-\cos\theta)/(n+a-1)$. The bound $(1-\cos\theta)/(n+a-1)$ is clearly better than $2/(n+a-1)$ obtained in Theorem 7. By recalling another result of this chapter, we may state here an improved version of Theorem 7, as

COROLLARY 4. *For real* $a > -n+1$ *and* $b > 0$, *the zeros* $\alpha_k^{(n)}(a,b)$ *of* $y_n(z;a,b)$ *satisfy the inequalities*

$$\frac{b}{n(n+a-1)} \le |\alpha_k^{(n)}(a,b)| \le \frac{1-\cos\theta}{n+a-1}\, b;$$

and the zeros $\beta_k^{(n)}(a,b)$ *of* $\theta_n(z;a,b)$ *satisfy*

$$\frac{n+a-1}{(1-\cos\theta)b} \le |\beta_k^{(n)}(a,b)| \le \frac{n(n+a-1)}{b}\,.$$

Under certain conditions we know that $\mathrm{Re}\,\alpha_k^{(n)}(a,b) < 0$. This happens when $a = 2$ (see Theorem 5), also for real $a \ne 2$, sufficiently large (see [74]) and, presumably, for all real a. Whenever that is the case, also $\mathrm{Re}\,\beta_k^{(n)}(a,b) < 0$ and θ in above inequalities may be restricted to $|\pi-\theta| < \pi/2$.

12. During the years 1956-59, Parodi perfected an approach for the determination of bounds for the characteristic values (eigenvalues) of matrices. It is based on the following classical theorem of Gershgorin, a proof of which may be found in [45].

THEOREM G. *Let* A *be an* $n \times n$ *matrix with complex coefficients* a_{jk}. *All characteristic values of* A *lie in the union* D_1 *of the disks* $|x-a_{jj}| \le \sum_{\substack{k=1\\k\ne j}}^{n} |a_{jk}|$. *They also lie in the union* D_2 *of the disks* $|x-a_{jj}| \le \sum_{\substack{j=1\\j\ne k}}^{n} |a_{jk}|$; *hence they lie in the intersection* $D_1 \cap D_2$.

In connection with the representation of BP by determinants (see Chapter 3), Parodi's method leads to bounds for the zeros of $y_n(z)$. These are weaker than those given by Theorem 7, but the method is simple, elegant, and may well be improved further. For these reasons we state here

THEOREM 8. *The zeros* α_k *of* $y_n(z)$ *belong to the intersection of the regions* D_1 *and* D_2, *defined by* D_1 = *union of the circles* $|z| \le 2/3$ *and* $|1+z| \le 1$; D_2 = *union of the circles* $|z| \le 6/5$ *and* $|1+z| \le 1/3$.

Proof. Let M_n be the determinant defined in Chapter 3. Then the zeros of $y_n(z)$ are the same as those of M_n. On the other hand $M_n = |A-Iz|$, where I is the $n \times n$ unit matrix and

$$A = \begin{vmatrix} 0 & -\frac{1}{2n-1} & 0 & \cdots & 0 & 0 & 0 \\ \frac{1}{2n-3} & 0 & -\frac{1}{2n-3} & \cdots & 0 & 0 & 0 \\ 0 & \frac{1}{2n-5} & 0 & \cdots & 0 & 0 & 0 \\ 0 & 0 & \frac{1}{2n-7} & \cdots & 0 & 0 & 0 \\ \vdots & \vdots & \vdots & \cdots & \vdots & \vdots & \vdots \\ 0 & 0 & 0 & \cdots & \frac{1}{3} & 0 & -\frac{1}{3} \\ 0 & 0 & 0 & \cdots & 0 & 1 & -1 \end{vmatrix}.$$

It follows that the zeros of $y_n(z)$ are precisely the characteristic values of A. These are located, by Theorem G and looking at rows, in the union of the circles $|z| \le \frac{1}{2n-1}$, $|z| \le \frac{2}{2n-2k+1}$ $(k = 2,3,\ldots,n-1)$ and $|z+1| \le 1$. Most circles are concentric and their union is D_1. In the same way, proceeding by columns, we obtain D_2 and the Theorem is proved. It is obvious that it can be improved, by adding the condition of Theorem 5 that Re $\alpha < 0$. Parodi shows in the same paper [85], how previous result can be sharpened. If we multiply the next to last column of M_n by $1+z$ and add to the last and expand the resulting determinant by the last line, we see that, with an irrelevant (but easily found) non-vanishing constant C_n, $y_n(z) = C_n|B-Iz|$, where B is the $(n-1) \times (n-1)$ matrix

$$B = \begin{vmatrix} 0 & -\frac{1}{2n-1} & 0 & \cdots & 0 & 0 & 0 \\ \frac{1}{2n-3} & 0 & -\frac{1}{2n-3} & \cdots & 0 & 0 & 0 \\ 0 & \frac{1}{2n-5} & 0 & \cdots & 0 & 0 & 0 \\ 0 & 0 & \frac{1}{2n-7} & \cdots & 0 & 0 & 0 \\ \vdots & \vdots & \vdots & \cdots & \vdots & \vdots & \vdots \\ 0 & 0 & 0 & \cdots & 0 & -\frac{1}{7} & 0 \\ 0 & 0 & 0 & \cdots & \frac{1}{5} & 0 & -\frac{1}{5}(1+z) \\ 0 & 0 & 0 & \cdots & 0 & \frac{1}{3} & -\frac{1}{3}-z(1+z) \end{vmatrix}.$$

By applying Gershgorin's Theorem G we now find that the zeros of $|B-Iz|$, hence, those of $y_n(z)$, belong to the intersection of the regions D_1' and D_2', where D_1' = union of $|z| \le \frac{1}{7}$, $|z| \le \frac{1}{5} + \frac{1}{5}|1+z|$, and $|\frac{1}{3} + z(1+z)| \le \frac{1}{3}$; D_2' = union of $|z| \le \frac{1}{3} + \frac{1}{7}$ and $|\frac{1}{3} + z(1+z)| \le \frac{1}{5}|1+z|$. A sketch shows immediately, that $D_1' \cap D_2'$ is much smaller than and contained in $D_1 \cap D_2$. The result can be improved further, by removing the portions of $D_1 \cap D_2$, or of $D_1' \cap D_2'$ in the right half plane.

13. A more powerful method to improve Parodi's approach is due to Wragg and Underhill [113]. They use, in addition to Gershgorin's theorem also the following Theorem H, due to Bendixson [3] and to Hirsch [33]. For a more recent presentation and proof, see [45].

THEOREM H. *Let* A *be an* n × n *matrix with complex entries, denote by* A^* *its complex conjugate and by* λ_j $(j = 1,2,\ldots,\nu)$ *is characteristic values. Set* $A = B + iC$, *where* $B = \frac{1}{2}(A+A^*)$ *and* $C = \frac{1}{2i}(A-A^*)$ *are Hermitian matrices and denote their respective characteristic values by* μ_j *and* ν_j, *respectively* $(j = 1,2,\ldots,n)$. *Then*

$$\min_j \mu_j \le \operatorname{Re} \lambda_k \le \operatorname*{Max}_j \mu_j \quad \textit{and} \quad \min_j \nu_j \le \operatorname{Im} \lambda_k \le \operatorname*{Max}_j \nu_j \quad \textit{for } k = 1,2,\ldots,n.$$

THEOREM 9. *If* $\alpha = \xi + i\eta$ *is a zero of* $y_n(z)$, *then* $-\frac{2}{3} \le \xi \le -\frac{2}{(2n-3)(2n-1)} < 0$ *and* $|\eta| \le 8/15$.

Proof. Similar matrices have the same characteristic values. Let $W = P^{-1}AP$, with A as previously defined and $P = ||p_{ij}||$, an n × n matrix with $p_{ij} = 1$ if $i+j \ge n+1$, $p_{ij} = 0$ otherwise. Then $P^{-1} = ||p_{ij}'||$ with $p_{ij}' = 1$ if $i+j = n+1$, $p_{ij}' = -1$ if $i+j = n$, $p_{ij}' = 0$ otherwise. One verifies that

$$W = \begin{bmatrix} -\frac{2}{3} & \frac{1}{3} & 0 & 0 & \cdots & 0 & 0 & 0 \\ -\frac{1}{3} & -\frac{2}{15} & \frac{1}{5} & 0 & \cdots & 0 & 0 & 0 \\ 0 & -\frac{1}{5} & -\frac{2}{35} & \frac{1}{7} & \cdots & 0 & 0 & 0 \\ \vdots & \vdots & \vdots & \vdots & \cdots & \vdots & \vdots & \vdots \\ 0 & 0 & 0 & 0 & \cdots & \frac{-1}{2n-3} & \frac{-2}{(2n-3)(2n-1)} & \frac{1}{2n-1} \\ 0 & 0 & 0 & 0 & \cdots & 0 & \frac{-1}{2n-1} & \frac{-1}{2n-1} \end{bmatrix}.$$

We now apply Theorem H, by writing $W = B+iC$, with B the diagonal matrix of the diagonal elements of W, and

$$C = \begin{bmatrix} 0 & -\frac{i}{3} & 0 & \cdots & 0 & 0 & 0 \\ \frac{i}{3} & 0 & -\frac{i}{5} & \cdots & 0 & 0 & 0 \\ \vdots & \vdots & \vdots & \cdots & \vdots & \vdots & \vdots \\ 0 & 0 & 0 & \cdots & \frac{i}{2n-3} & 0 & \frac{-i}{2n-1} \\ 0 & 0 & 0 & \cdots & 0 & \frac{i}{2n-1} & 0 \end{bmatrix}.$$

The characteristic values of B are the diagonal entries themselves, so that, by Theorem H, $-\frac{2}{3} \le \xi \le -\frac{2}{(2n-3)(2n-1)}$. By Theorem G, the characteristic values of C lie in the union of the circles $|z| \le \frac{1}{3}$, $|z| \le \frac{1}{2k-3} + \frac{1}{2k-1}$ $(k = 3,4,\dots,n)$ and $|z| \le \frac{1}{2n-1}$; they are concentric and the largest corresponds to $k = 3$, whence the theorem.

14. In 1952 and 1954 appeared two important papers of F.W.J. Olver (see [84] and [46]). In them the author discusses expansions of functions that satisfy a certain class of differential equations and studies their zeros. In particular, he puts special emphasis on Bessel functions. Specifically, he determined, ([84], p. 353) among others the distribution of zeros of the Hankel functions $H_\nu^{(1)}(z)$ and $H_\nu^{(2)}(z)$, for integral and for half-integral ν. From these the zeros of $K_\nu(z)$ follow immediately because (see [1], 9.6.4).

$$K_\nu(z) = \begin{cases} = \frac{1}{2}\pi i e^{\frac{1}{2}\nu\pi i} H_\nu^{(1)}(ze^{\frac{\pi}{2}i}) (-\pi < \arg z < \pi/2) \\ = -\frac{1}{2}\pi i e^{-\frac{1}{2}\nu\pi i} H_\nu^{(2)}(ze^{-\frac{\pi i}{2}}) (-\frac{\pi}{2} < \arg z < \pi). \end{cases}$$

For half integral values of $\nu = n + 1/2$, the zeros of $K_\nu(z)$ are (see Chapter 2) of course the same as the $\beta_k^{(n)}$, that is the reciprocals of the $\alpha_k^{(n)}$. The theory of Olver and even his precise conclusions cannot be presented here and the interested reader is advised to consult Olver's original papers [84], [**46**]. A simplified version of his results, however, is easy to state. He shows that the zeros of $H^{(1)}_{n+1/2}(ze^{\frac{\pi}{2}i})$ lie on a convex arc (see Fig. 15 on p. 352 of [84]), symmetric with respect to the imaginary axis. For $n \to \infty$, the points of intersection with the real axis approach $-n$ and $+n$, while that with the imaginary axis approaches inz_0, where $z_0 = \sqrt{t_0^2 - 1} \cong .66274\ldots$. Here $t_0 \cong 1.19967\ldots$, is the (unique) positive root of the equation $\coth t = t$.

The zeros of $K_{n+1/2}(z)$, i.e. of $\theta_n(z)$, are then located (see e.g. [**1**], p. 377 and Fig. 9.6) on an arc obtained by rotating the previous arc by $\pi/2$. Finally, the zeros of $y_n(z)$ are on an arc obtained from the preceding one by inversion in the unit circle. (see Fig. 1). Let $v = z_0^{-1} \cong 1.50888\ldots$

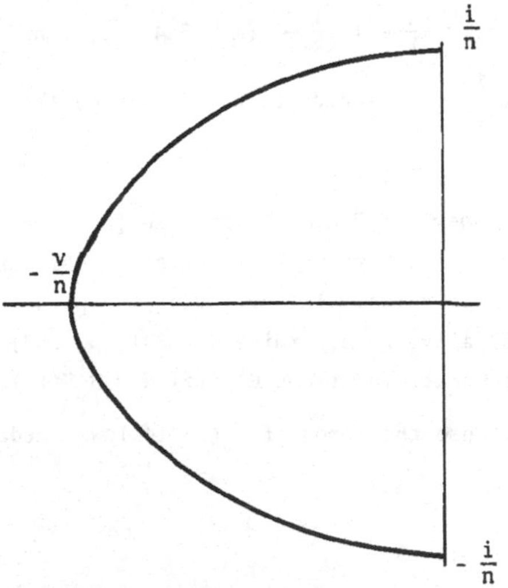

Figure 1

then the results of Olver have as a corollary the following statements.

THEOREM 10. *All the zeros* $\alpha_k^{(n)}$ *of* $y_n(z)$ *are located on an arc symmetric with respect to the real axis, that runs in the left half-place. For* $n \to \infty$, *the intersection of that arc with the imaginary axis approach* $\pm i/n$, *while that with the real axis approaches* $-v/n$ (see Fig. 1). *The absolute largest zero of the polynomials* $y_n(z)$ *of odd degree is the real one, say* $\alpha_m^{(n)}$ $(m = (n+1)/2)$ *and* $\alpha_m^{(n)} = -vn^{-1} + O(n^{-2})$.

To give an idea of the degree of approximation in Olver's results, we may consider the following tabulation:

n	$-\nu/n$	$\alpha_m^{(n)}$
1	-1.50888...	-1
3	- .5029...	- .4305...
5	- .3017...	- .2742...

While Olver's results contain an element of finality, they yield exact results only in the limit, as $n \to \infty$.

Numerical investigations have led to the following

<u>CONJECTURE</u>. For arbitrary $a > 0$, $b = 2$ and odd $n \to \infty$, the real zero $\alpha_m^{(n)}(a,2)$ of $y_n(z;a,2)$ satisfies the asymptotic equality

$$\alpha_m^{(n)}(a,2) \sim -2\{1.32548\, n + a - (\pi+1)/\pi\}^{-1}.$$

This follows conjecturally from [71], vol. 2, p. 194, by taking into account the differences of notation and normalization.

15. Burchnall's results (2), (2') turned out to be useful in some rather unexpected applications (see, e.g. [62], [55]). In fact, it would be very desirable to have simple explicit formulae for the sum of arbitrary powers of the zeros of $\theta_n(z)$, or $y_n(z)$. No such formulae, valid for all powers seem to be known.

A certain amoung of information about these sums, can, of course, be obtained from Newton's classical formulae (see, e.g. [30], p. 208). Let us denote the sums of the r-th power of the zeros of a polynomial by σ_r and, if we want to put into evidence that the polynomial is $y_n(z)$, then we write $\sigma_r^{(n)}$. For a general polynomial $f(x) = \sum_{k=0}^{n} c_k x^{n-k}$ with $c_0 = 1$ Newton's identities are

$$\sum_{j=0}^{r-1} c_j\, \sigma_{r-j} + r c_r = 0 \quad \text{for } r = 1,2,\ldots,n;$$

(13)

$$\sum_{j=0}^{n} c_j\, \sigma_{r-j} = 0 \qquad \text{for } r > n.$$

In the case of $y_n(z;a)$, we may take $c_j = d_{n-j}^{(n)}/d_n^{(n)}$, with the $d_k^{(n)}$ given by (2.27). This leads to $c_1 = \frac{2n}{2n+a-2}$, $c_2 = \binom{n}{2}\frac{2^2}{(2n+a-2)(2n+a-3)}, \ldots, c_j = \binom{n}{j}\frac{2^j}{(2n+a-2)^{(j)}}, \ldots,$ $c_n = \frac{2^n}{(2n+a-2)^{(n)}} = 2^n \frac{(n+a-2)!}{(2n+a-2)!}$. By (13) (assuming n large) we obtain $\sigma_1 = -c_1 = \frac{-2n}{2n+a-2}$, $\sigma_2 = -\sigma_1 c_1 - 2c_2 = c_1^2 - 2c_2 = \frac{4n(n+a-2)}{(2n+a-2)^2(2n+a-3)}$,

$\sigma_3 = \frac{-8(a-2)(n+a-2)n}{(2n+a-2)^3(2n+a-3)(2n+a-4)}$, etc. It is immediately apparent that we obtain significant simplifications, if we restrict ourselves to the case a = 2, of the simple BP. In that case $c_j = \binom{n}{j} \frac{2^j}{(2n)^{(j)}}$ and previous formulae reduce to $\sigma_1 = -1$, $\sigma_2 = \frac{1}{2n-1}$, $\sigma_3 = 0$; also, $\sigma_4 = \frac{-1}{(2n-1)^2(2n-3)}$, $\sigma_5 = 0$, $\sigma_6 = \frac{2}{(2n-1)^3(2n-3)(2n-5)}$, $\sigma_7 = 0$, $\sigma_8 = -\frac{10n-17}{(2n-1)^4(2n-3)^2(2n-5)(2n-7)}$, $\sigma_9 = 0$, etc.

The values obtained for odd subscripts are, of course, those predicted by Burchnall's theorem (2).

The sums of negative powers of the zeros $\alpha_k^{(n)}$ of $y_n(z)$ are the same as the sums of the positive powers of the zeros $\beta_k^{(n)}$ of $\theta_n(z)$ and conversely. In order to avoid confusion, in what follows we shall denote by $\sigma_r = \sigma_r^{(n)}$ the sum of the r-th powers (r positive, zero, or negative) of the zeros $\alpha_k^{(n)}$ of the simple BP $y_n(z)$. The sums σ_{-r} of negative powers can of course be computed as sums of positive powers of the zeros $\beta_k^{(n)}$ of $\theta_n(z)$, by use of Newton's formulae (13). The coefficients c_j in (13) are now precisely the a_j's of (2.8).

With $a_0 = 1$, $a_1 = \frac{(n+1)n}{2}$, $a_2 = \frac{(n+2)^{(4)}}{2^2 2!}$, ... we obtain $\sigma_{-1} = -a_1 = -\frac{(n+1)n}{2}$, $\sigma_2 = -a\sigma_{-1}-2a_2 = a_1^2-2a_2 = \frac{(n+1)n}{2}$, $\sigma_{-3} = \frac{1}{8}n(n+1)\{n(n+1)-6\}$, $\sigma_{-4} = -\frac{1}{2}n(n+1)\{n(n+1)-3\}$, $\sigma_{-5} = -\frac{1}{16}n(n+1)\{n^2(n+1)^2-28n(n+1)+60\}$ etc.

It turns out that we obtain somewhat simpler results if we set $n(n+1)/2 = N$, say. Then $\sigma_{-1} = -N$, $\sigma_{-2} = \frac{1}{2}N(N+1)$, $\sigma_{-3} = \frac{1}{2}N(N-3)$, $\sigma_{-4} = -N(2N-3)$, $\sigma_{-5} = -\frac{1}{2}N(N^2-14N+15)$, etc.

In this way we are able to compute effectively the sums of any given power (positive or negative) of the zeros of the BP $y_n(z)$; however, no obvious pattern seems to emerge and an explicit formula for $\sigma_r^{(n)}$ and/or $\sigma_{-r}^{(n)}$ does not seem to be known.

A more convenient way to approach this problem is through a recursion formula (see [62]), valid at least for positive values of r. While of a rather complicated appearance, it permits one to go beyond Burchnall's (2), for which incidentally it gives a new proof.

We start by observing that, on account of Theorem 1 all zeros of BP are simple. Hence,

$$\frac{\theta_{n-1}(z)}{\theta_n(z)} = \sum_{j=1}^{n} \frac{A_{nj}}{z-\beta_j}, \text{ with } A_{nj} = \lim_{z \to \beta_j} (z-\beta_j) \frac{\theta_{n-1}(z)}{\theta_n(z)} = \frac{\theta_{n-1}(\beta_j)}{\theta'_n(\beta_j)} .$$

By (3.8) the last fraction equals $-\beta_j^{-1}$, so that

$$(14) \qquad \frac{\theta_{n-1}(z)}{\theta_n(z)} = \sum_{j=1}^{n} \frac{1}{\beta_j(\beta_j - z)} .$$

The corresponding formula for $y_n(z)$ is easily obtained from (14), by replacing $\theta_n(z)$ by $z^n y_n(1/z)$, β_j by α_j^{-1} and then simplifying the result. We obtain

$$\frac{y_{n-1}(z)}{y_n(z)} = \sum_{j=1}^{n} \frac{\alpha_j^2}{z-\alpha_j} .$$

Returning to (14), we observe that it can be written as

$$(15) \quad \frac{\theta_{n-1}(z)}{\theta_n(z)} = \sum_{j=1}^{n} \frac{1}{\beta_j^2(1-z/\beta_j)} = \sum_{j=1}^{n} \beta_j^{-2} \sum_{r=0}^{\infty} \left(\frac{z}{\beta_j}\right)^r = \sum_{r=0}^{\infty} z^r \sum_{j=1}^{n} \beta_j^{-r-2} = \sum_{r=0}^{\infty} \sigma_{r+2}^{(n)} z^r .$$

On the other hand, by (3.5) (with n replaced by n-1),

$$\frac{\theta_n(z)}{\theta_{n-1}(z)} = 2n-1+z^2 \frac{\theta_{n-2}(z)}{\theta_{n-1}(z)} \qquad \text{or}$$

$$(16) \quad \frac{\theta_{n-1}(z)}{\theta_n(z)} = \frac{1}{2n-1+z^2\theta_{n-2}(z)/\theta_{n-1}(z)} = \frac{1}{2n-1} \sum_{m=0}^{\infty} (-1)^m \left(\frac{z^2}{2n-1} \frac{\theta_{n-2}(z)}{\theta_{n-1}(z)}\right)^m .$$

We now equate the right hand sides of (15) and of (16) and then replace the ratios $\theta_{n-2}(z)/\theta_{n-1}(z)$ with the help of (15). We obtain

$$\sum_{r=0}^{\infty} z^r \sigma_{r+2}^{(n)} = \sum_{m=0}^{\infty} (-1)^m \frac{z^{2m}}{(2n-1)^{2m+1}} \{ \sum_{k=0}^{\infty} z^k \sigma_{k+2}^{(n-1)} \}^m$$

$$= \frac{1}{2n-1} + \sum_{m=1}^{\infty} \frac{(-1)^m z^{2m}}{(2n-1)^{m+1}} \left(\sum_{k=0}^{\infty} z^k \sigma_{k+2}^{(n-1)} \right)^m$$

$$= \frac{1}{2n-1} + \sum_{r=1}^{\infty} z^r \sum_{\substack{2m+k_1+\ldots+k_m=r \\ m \geq 1,\ k_j \geq 0}} \frac{(-1)^m}{(2n-1)^{m+1}} \sigma_{k_1+2}^{(n-1)} \sigma_{k_2+2}^{(n-1)} \cdots \sigma_{k_m+2}^{(n-1)} .$$

By comparing coefficients, we obtain first $\sigma_2^{(n)} = \frac{1}{2n-1}$, then, in general, for $r > 0$,

$$(17)\qquad \sigma_{r+2}^{(n)} = \sum_{\substack{2m+\sum_{j=1}^{m} k_j = r \\ m\geq 1,\ k_j \geq 0}} \frac{(-1)^m}{(2n-1)^{m+1}} \sigma_{k_1+2}^{(n-1)} \cdots \sigma_{k_m+2}^{(n-1)} .$$

Formula (17) permits one, in principle, to compute the sums $\sigma_m^{(n)}$ recursively, if one already knows the sums $\sigma_m^{(n-1)}$. In fact, (17) looks rather forbidding and one may wonder whether it can be put to any use. Fortunately, the answer is positive, although (so far, at least) (17) could not be made to yield explicit formulae for all $\sigma_m^{(n)}$.

To start with, we may use (17) in order to obtain a new proof of (2), (2'). We observe that for n = 1 and n = 2 we have

$$\frac{\theta_0(z)}{\theta_1(z)} = \frac{1}{1+z} = \sum_{r=0}^{\infty} (-1)^r z^r = \sum_{r=0}^{\infty} \sigma_{r+2}^{(1)} z^r$$

and

$$\frac{\theta_1(z)}{\theta_2(z)} = \frac{1+z}{3+3z+z^2} = \frac{1}{3}\,\frac{1+z}{1+z+z^2/3} = \frac{1}{3}\,\frac{1}{1+z^2/3(1+z)} =$$

$$= \frac{1}{3}\{1-\frac{z^2}{3}(1-z+z^2-z^3+\ldots) + \frac{z^4}{9}(1-z+z^2-z^3+\ldots)^2 - \ldots\}$$

$$= \frac{1}{3} - \frac{z^2}{9} + \frac{z^3}{9} - \frac{2}{27}z^4 + \frac{1}{27}z^5 + \ldots = \sum_{r=0}^{\infty} \sigma_{r+2}^{(2)} z^r .$$

We note that $\sigma_3^{(2)} = 0$; this is the instance n = 2 of the claim that $\sigma_{2\nu+1}^{(n)} = 0$ holds for $\nu = 1,2,\ldots,n-1$. (For n = 1 this is also true - vacuously.) Next $\sigma_3^{(1)} = \sigma_5^{(1)} = -1$ and $\sigma_5^{(2)} = \frac{1}{9}$, $\sigma_7^{(2)} = \frac{1}{27}$, so that

$$(18)\qquad \sigma_{2n+1}^{(n)} = \frac{(-1)^n}{\{1.3.\ldots.(2n-1)\}^2}, \quad \sigma_{2n+3}^{(n)} = \frac{(-1)^n}{\{1.3\ldots(2n-1)\}^2(2n-1)}$$

both hold for n = 1 and n = 2. Let us assume that $\sigma_{2\nu+1}^{(k)} = 0$ has already been verified for all k with $2 \leq k \leq n-1$ and all ν, $1 \leq \nu \leq k-1$. Then (17) shows that if $r+2 = 2\nu+1$ is odd, also $\sigma_{r+2}^{(n)} = \sigma_{2\nu+1}^{(n)} = 0$ for $\nu = 1,2,\ldots,n-1$. Indeed, let k = n-1; if r+2 is odd, then at least one of the k_j's is odd. Also, all odd k_j's satisfy $1 \leq k_j \leq r-2m = 2\nu-1-2m \leq 2\nu-3 \leq 2n-5$, for $\nu \leq n-1$. In particular, we find in each product $\sigma_{k_1+2}^{(n-1)} \ldots \sigma_{k_m+2}^{(n-1)}$ at least one factor $\sigma_{k_j+2}^{(n-1)}$ with $3 \leq k_j+2 \leq 2n-3 = 2(n-1)-1$, and this vanishes by the induction hypothesis. It follows that all summands in (17) vanish and $\sigma^{(n)}_{2\nu+1} = 0$ for all $\nu = 1,2,\ldots,n-1$ in agreement with (2) and that is what we wanted to prove.

If $r+2 = 2n+1$, so that $r = 2n-1$, then one of the summands in (17) has $m = 1$ and $k_1 = r-2 = 2n-3$ (no other k_j's occur) and does not vanish. Indeed, (17) now reduces to $\sigma^{(n)}_{2n+1} = -\dfrac{\sigma^{(n-1)}_{2n-1}}{(2n-1)^2}$. The numerator is $\sigma^{(n-1)}_{2n-1} = \sigma^{(n-1)}_{2(n-1)+1}$, and we have verified that (18) holds at least for the superscripts 1 and 2; if we assume that the first of (18) has already been recognized as valid up to the superscript n-1, then $\sigma^{(n)}_{2n+1} = -\dfrac{1}{(2n-1)^2}\cdot\dfrac{(-1)^{n-1}}{\{1.3\ldots(2n-3)\}^2} = \dfrac{(-1)^n}{\{1.3\ldots(2n-1)\}^2}$ and (18) holds also for superscripts n, and this concludes its proof by induction.

If $r+2 = 2n+3$, then there are three potentially non-vanishing terms in (17), corresponding to $m = 1$, $k_1 = 2n-1$; $m = 2$, $k_1 = 2n-3$, $k_2 = 0$; and $m = 2$, $k_1 = 0$, $k_2 = 2n-3$. Hence, assuming that the second of (18) has been verified for $\sigma^{(k)}_{2k+3}$ and k up to n-1, (17) and the induction hypothesis yield

$$\sigma^{(n)}_{2n+3} = -\frac{\sigma^{(n-1)}_{2n+1}}{(2n-1)^2} + 2\,\frac{\sigma^{(n-1)}_{2}\sigma^{(n-1)}_{2n-1}}{(2n-1)^3} = \frac{(-1)^n}{\{1.3\ldots(2n-3)\}^2(2n-3)(2n-1)^2} +$$

$$2\cdot\frac{1}{2n-3}\,\frac{(-1)^{n-1}}{\{1.3\ldots(2n-3)\}^2}\,\frac{1}{(2n-1)^3} = \frac{(-1)^n}{\{1.3\ldots(2n-1)\}^2}\left\{\frac{1}{2n-3} - \frac{2}{(2n-1)(2n-3)}\right\}$$

$$= \frac{(-1)^n}{\{1.3\ldots(2n-1)\}^2(2n-1)}$$

and this finishes the proof by induction of (18).

One may, of course, go on. However, the number of non-vanishing terms in (17) increases fast and the formulae become rather complicated.

We shall finish this section with the remark that, although we do not know explicit formulae for $\sigma^{(n)}_r$, valid for all integral values of r and n, we still can at least prove the following

<u>THEOREM 11.</u> *For* $n = 1$, $\sigma^{(1)}_r = (-1)^r$. *For* $n = 2$, $\sigma^{(2)}_m = 3^{-[m/2]}h_m$ *and* $\sigma^{(2)}_{-m} = 3^{[(m+1)/2]}h_m$, *where* h_m *depends only on the residue class of* m *modulo* 12, *as follows:*

m	0	1	2	3	4	5	6	7	8	9	10	11	modulo 12
h_m	2	-1	1	0	-1	1	-2	1	-1	0	1	-1	

<u>Proof.</u> For $n = 1$, the result is obvious. For $n = 2$ we obtain by Newton's first

formula (13), that $\sigma_0 = 2$, $\sigma_1 = -1$, $\sigma_2 = 1/3$. Next, by the second formula (13), we verify that also $\sigma_3 = 0$, $\sigma_4 = -1/9, \dots, \sigma_{11} = -3^{-5}$, conform to the statement of the theorem. In general, by (13),

$$\sigma_m = -\sigma_{m-1} - \frac{1}{3}\sigma_{m-2}. \tag{19}$$

We assume that for all values of $k \leq m-1$ we have already verified that Theorem 11 holds. Then we check by (19) that also σ_m takes on the value prescribed by the theorem. This requires 12 checks, according to the residue class of m (mod 12). We may reduce these to 6 checks, by observing that $h_{m+6} = -h_m$. As an example, we work out one of these checks, say for $m \equiv 5 \pmod{12}$. Then $m-1 \equiv 4$ and $m-2 \equiv 3$ and, by the induction hypothesis and (13),

$$\sigma_m = -\sigma_{m-1} - \frac{1}{3}\sigma_{m-2} = -3^{[(m-1)/2]}h_4 - \frac{1}{3}\cdot 3^{-[(m-2)/2]}h_3 = 3^{-(m-1)/2} = 3^{-[m/2]}$$

and this proves that Theorem 11 holds also for m, provided that $m \equiv 5 \pmod{12}$, and in fact, also for $m \equiv 11 \pmod{12}$.

The assertion concerning $\sigma^{(2)}_{-m}$ could be proved in the same way, by using Newton's formulae for $\theta_2(z) = z^2+3z+3$. It is much easier, however, to observe that $\sigma_{-m} = \alpha_1^{-m} + \alpha_2^{-m} = (\alpha_1^m + \alpha_2^m)/(\alpha_1\alpha_2)^m = 3^m\sigma_m = 3^{m-[m/2]}h_m = 3^{[(m+1)/2]}h_m$, as claimed.

No similarly simple results seem to hold for $\sigma^{(n)}_m$, even when $n = 3$.

CHAPTER 11

ON THE ALGEBRAIC IRREDUCIBILITY OF THE BP

1. Much of the present chapter covers material found in [53]. However, in [53], there are many errors of detail and the presentation leaves much to be desired. In the present chapter many procfs have been reworked completely, in order to make the treatment not only correct, but also readable.

The term *irreducible* will mean here irreducible over the field of rational numbers. This is equivalent to the statement that if f,g and h are polynomials with rational, integral coefficients, then f = gh is possible only if either g, or h reduce to a constant polynomial, in fact to a rational integer, and the other factor has the same degree as f. The polynomials here considered will, in fact, be monic.

It is extremely likely that all BP are irreducible, but no proof of this general statement exists. What we can prove are the four theorems of the present chapter. The symbol p, with or without subscripts, stands for an odd prime, while q and m are positive inters. Also, for given n, we set $k_1 = \min_{p_1 \le n} (n-p_1)$, $k_2 = \min_{p_2 > n} (p_2-n-1)$, and $k = k(n) = \min (k_1,k_2)$.

THEOREM 1. (a) *The* BP *of degrees* $n = p^m$ *are irreducible.*

(b) *The* BP *of degrees* $n = qp-1$, *with* $q < p/2$ *are either irreducible, or else can have only irreducible factors of degrees* rp *or* $rp-1$ $(r = 1,2,\dots,q)$. *If* $p-1 > k(n)$, *then the* BP *is irreducible. In particular, the* BP *of degrees* $n = p-1$ *are irreducible.*

(c) *The* BP *of degrees* $n = qp+1$, *with* $q < p/2$ *are either irreducible, or else can have only irreducible factors of degrees* rp, *or* $rp+1$ $(r = 0,1,\dots, q)$. *If also* $p > k(n)$ *and* q *is odd, then the* BP *is irreducible. In particular, the* BP *of degrees* $n = p+1$ *are irreducible.*

(d) *The* BP *of degrees* $n = qp$ *with* $q < p/2$ *can have irreducible factors only of degrees that are multiples of* p. *If also either* $p > k(n)$, *or* $q \le 17$, *then the* BP *is irreducible.*

(e) *The* BP *of degrees* $n = p^k-1$ *can only have irreducible factors with degrees that are multiples of* $p-1$. *If also* $p-1 > k(n)$, *holds, then the* BP *is irreducible.*

(f) *If* p *is the largest prime factor of* n, *or of* n+1, *then the* BP *of degree* n *cannot have any factors of degree less than* p-1.

THEOREM 2. *For every integer* n, *the* BP *of degree* n *contains an irreducible factor of degree* $A_n n$, *with* $\lim_{n \to \infty} A_n = 1$. *For every* n, $A_n > 16/17$ *and no* BP *can have an irreducible factor of degree* d, *with* $n/17 < d < (16/17)n$.

THEOREM 3. *All* BP *of degree* $n \le 400$ *are irreducible.*

There is no doubt that with more work the bound $n \le 400$ could be increased. In fact, there are only four values of n, $301 \le n \le 400$ for which a separate verification was needed. Also the fraction 1/17 that occurs in Theorem 2 could be decreased and Theorem 1 could be improved by the addition of other types of n, for which irreducibility can be guaranteed. There does not seem, however, to be much point to such a task, because in any case it would fall short of the proof of the already mentioned

CONJECTURE 1. *All* BP *are irreducible.*

2. The proofs of these theorems rely heavily on Theorem A, due to Dumas [12], to be stated presently. It is based on the theory of the Newton polygon of a polynomial f(x), with rational, integral coefficients, with respect to a fixed, rational prime p (see [12] and [28]). This frequently occurring sentence will be abbreviated to read "N.p. of f(x), w.r. to p", or simply "N.p.".

DEFINITION. *Let* $f(x) = \sum_{m=0}^{n} a_m p^{e_m} x^{n-m}$, a_m, e_m *rational integers*, $p \nmid a_m$, *and consider the points** $P_m = [m, e_m]$. *For want of a better name we call these particular lattice points of the plane the* spots *of* f(x). *The* N.p. of f(x) w.r. *to* p *is the unique open polygonal line with vertices only at the spots of* f(x), *convex downwards and with no spots below its sides* (see Fig. 1).

THEOREM A. (Dumas [12]). *Let* $f(x) = \sum_{m=0}^{n} a_m p^{e_m} x^{n-m}$, $p \nmid a_m$; *let* h_r *and* v_r *be the horizontal and vertical projections, respectively, of the* r*-th side of the N.p. of* f(x) w.r. *to* p *and let their greatest common divisor be* $t_r = (h_r, v_r)$. *Then, if* M *is the number of sides of the N.p. and* $h_r = t_r s_r$, *all factors of* f(x) *have degrees of the form* $\sum_{r=1}^{M} \mu_r s_r$, *with integers* μ_r, $0 \le \mu_r \le t_r$.

REMARK. If $v_r = 0$, then $t_r = h_r$ and $s_r = 1$.

COROLLARY 1. *If* $e_0 = 0$, $e_m \ge m e_n / n$ *for* $m = 1, 2, \ldots, n$, *then* $M = 1$ *and all factors of* f(x) *have degrees of the form* $\mu n/t$, $t = (n, e_n)$, $1 \le \mu \le t$. *If, furthermore,* $(n, e_n) = 1$, *then* $\mu = t = 1$ *and* f(x) *is irreducible.*

Proof of the Corollary follows immediately from Theorem A.

In what follows it will be convenient to refer (improperly!) to the horizontal projection h_r, as the *length* of the r-th side of N.p. We shall not give here a proof of Theorem A. Indeed, an excellent presentation can be found in [59] (see also [12], or [60]), and none of the available proofs is really easy.

* We use square brackets rather than parentheses, in order to avoid confusion with the notation (m, e_m) used for the greatest common divisor of the integers m, e_m.

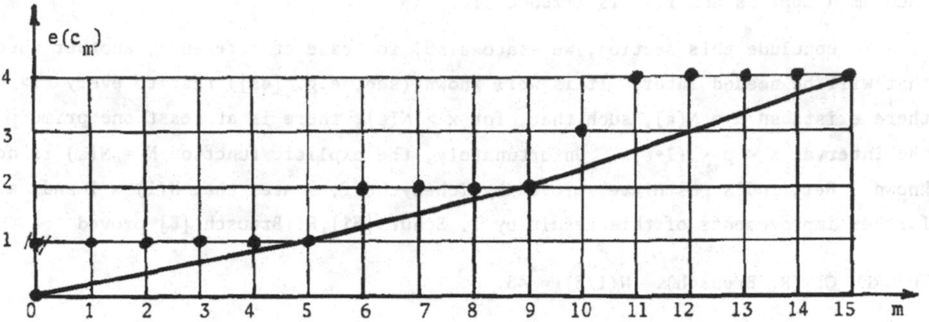

Newton Polygon of the polynomial $z_{15}(x) = z^{15}+5(c_1z^{14}+c_2z^{13}+c_3z^{12}+c_4z^{11}+c_5z^{10})$ $+5^2(c_6z^9+c_7z^8+c_8z^7+c_9z^6)+5^3c_{10}z^5+5^4(c_{11}z^4+c_{12}z^3+c_{13}z^2+c_{14}z+c_{15})$, $5 \nmid c_j$ $(j = 1,2,\ldots,15)$, modulo $p = 5$.

Figure 1.

Sometimes a particularly simple case of Theorem A is sufficient for the purpose on hand. This is known as Eisenstein's criterion of irreducibility.

THEOREM B. (Eisenstein; see [59]). *Let* $f(x) = x^n + p\sum_{m=1}^{n} a_m x^{n-m}$, *with* $p\nmid a_n$; *then* $f(x)$ *is irreducible.*

Proof. Here $e_0 = 0$, $e_n = 1$, so that the N.p. of $f(x)$ reduces to a single line of length n. As also $e_m \geq 1$ for $1 \leq m \leq n$ and $(n,e_n) = (n,1) = 1$, the Corollary to Theorem A applies and $f(x)$ is irreducible.

To conclude this section, we state also, for ease of reference, another theorem that will be needed later. It is well known (see, e.g. [43]) that to every $\varepsilon > 0$, there exists an $N = N(\varepsilon)$, such that, for $x \geq N(\varepsilon)$, there is at least one prime p in the interval $x < p \leq (1+\varepsilon)x$. Unfortunately, the explicit function $N = N(\varepsilon)$ is not known. Bertrand's postulate, proved by Tchebycheff, states that $N(1) = 1$ and, after further improvements of this result by I. Schur [51], R. Breusch [8] proved

THEOREM C. (R. Breusch). $N(1/8) = 48$.

3. Let us denote by $e(m)$ ($= e_p(m)$; we shall suppress the subscript) the exact power of p that divides m.

For any integer n, it is clear (see, e.g. [24]) that

$$e(n!) = \left[\frac{n}{p}\right] + \left[\frac{n}{p^2}\right] + \dots + \left[\frac{n}{p^r}\right] + \dots \tag{1}$$

where $[x]$ stands for the greatest integer not in excess of x; there should be no danger of confusion with $P_m = [m,e_m]$, where there are two arguments enclosed in square brackets. It is sufficient to stop in (1) with the term $r = \left[\frac{\log n}{\log p}\right]$, because all other terms vanish. For $n = p^m$, (1) immediately yields

$$e((p^m)!) = \frac{n-1}{p-1} . \tag{2}$$

More generally, let $n = \sum_{j=0}^{k} a_j p^j$, $0 \leq a_j \leq p-1$ be the p-adic representation of n; then we obtain by direct substitution in (1),

$$\begin{aligned} e(n!) = &+ a_k p^{k-1} + a_{k-1}p^{k-2} + \dots + a_2 p + a_1 \\ &+ a_k p^{k-2} + a_{k-1}p^{k-3} + \dots + a_2 \\ &+ \dots\dots \\ &+ a_k p \quad + a_{k-1} \\ &+ a_k \end{aligned}$$

$$= a_k(p^{k-1} + p^{k-2} + \ldots + p+1) + a_{k-1}(p^{k-2} + p^{k-3} + \ldots + 1) + \ldots + a_2(p+1) + a_1$$

$$= a_k \frac{p^k-1}{p-1} + a_{k-1} \frac{p^{k-1}-1}{p-1} + \ldots + a_1 \frac{p^2-1}{p-1}$$

$$= \frac{1}{p-1} \{ \sum_{j=0}^{k} a_j p^j - \sum_{j=0}^{k} a_j \}.$$

This completes the proof of the following lemma, due to Legendre:

LEMMA 1. $e(n!) = \frac{n-\sigma(n)}{p-1}$, *where* $\sigma(n)\ (= \sigma_p(n)) = \sum_{j=0}^{n} a_j$, $\sigma(0) = 0$.

From Lemma 1 immediately follows

LEMMA 2. *Set*

$$c_m = \frac{(n+m)!}{m!(n-m)!}; \tag{3}$$

then $e(c_m) = \frac{m+\sigma(m)+\sigma(n-m)-\sigma(n+m)}{p-1}$.

Proof. By Lemma 1, $e(c_m) = \frac{1}{p-1}\{n+m-\sigma(n+m)-m+\sigma(m)-(n-m)+\sigma(n-m)\}$ and Lemma 2 is proved

4. It is clear that $y_n(x)$ and $\theta_n(x)$ are either both reducible, or both irreducible. In fact, there is an obvious one-to-one correspondence between their respective irreducible factors. It is somewhat more convenient to apply previous criteria of irreducibility to monic polynomials and, therefore, we shall be concerned here with $\theta_n(x)$, rather than $y_n(x)$. We obtain a slight additional simplification, if we observe that $\theta_n(x/2) = 2^{-n} \sum_{m=0}^{n} \frac{(n+m)!}{m!(n-m)!} x^{n-m}$, so that, by using also (3), $z_n(x) = 2^n\theta_n(x/2) = \sum_{m=0}^{n} c_m x^{n-m}$. It is sufficient to study the irreducibility of $z_n(x)$ and, in this chapter only, conclusions referring to BP of degree n will apply to any of the three polynomials $y_n(x)$, $\theta_n(x)$ or $z_n(x)$.

Proof of Theorem 1. Part I. (i). Let $n = p^k$; then $2n = 2p^k$ and, by Lemma 1, or 2 $e(c_n) = \frac{1}{p-1}\{2n-2-n+1\} = \frac{n-1}{p-1} = \sum_{j=0}^{k-1} p^j$. Clearly, $(e(c_n),n) = 1$ holds, but we observe that this would no longer be the case for $p = 2$, $n = 2^m$, where $2n = 2^{m+1}$,

$e(c_n) = n$, so that $(e(c_n),n) = n$. This is the reason, why p is assumed to be an *odd* prime only.

We claim that, for $1 \le m \le n-1$,

$$e(c_m) > \frac{m}{n} e(c_n) = \frac{m(n-1)}{n(p-1)} . \tag{4}$$

Assuming for a moment (4), it follows that the N.p. of $z_n(x)$ reduces to the straight line from $P_0 = [0,0]$ to $P_n = [n, e(c_n)]$. On account of $(n,e(c_n)) = 1$, Corollary 1 shows that $z_n(x)$, is irreducible. It remains to prove (4).

Let $m = \sum_{j=0}^{k-1} a_j p^j$, $0 \le a_j \le p-1$, with $\sigma(m) = \sum_{j=0}^{k-1} a_j$. Then $n+m = p^k + \sum_{j=0}^{k-1} a_j p^j$,

$\sigma(n+m) = \sigma(m) + 1$ and, by Lemma 2,

$$e(c_m) = \frac{1}{p-1} (m+\sigma(m)+\sigma(n-m)-\sigma(n+m)) = \frac{1}{p-1} (m+\sigma(n-m)-1).$$

To prove (4), we have to verify that $n(m+\sigma(n-m)-1) > m(n-1)$, or that $n(\sigma(n-m)-1) > -m$ This is obviously true, because $\sigma(n-m)$ and m are both positive integers. Part (a) is proved.

(ii) We now dispose first of a few other easy to prove statements of Theorem A. If $n = p-1$, then $p|c_m$ for $m = 1,2,\dots,n$ and $p^2 \nmid c_n$ ($= p(p+1)\dots(2p-2)$), so that Theorem B applies and $z_n(x)$ is irreducible. If $n = p+1$, then $p|c_m$ for $2 \le m \le n$, $p^2 \nmid c_n$, because $c_n = (p+2)(p+3)\dots(2p)(2p+1)(2p+2)$. It follows that the N.p. of $z_n(x)$ consists of two sides, the first of length one, from $P_0 = [0,0]$ to $P_1 = [1,0]$, and the second from P_1 to $P_n = [n,1]$ of length $(n-1)$ and without spots on it. By Theorem A, $z_n(x)$ is either irreducible, or else splits into a linear factor and an irreducible polynomial of degree $n-1 = p$. The last alternative, however, is impossible. Indeed, the linear factor has a real zero, (which, in fact, must be a negative integer) and we know (see Chapter 10) that BP of even degree ($n = p+1$ is even) has no real zeros. It follows that $z_n(x)$ is irreducible. Let us observe at this point that we already have proved so far a weaker form of Theorem 3, namely

<u>THEOREM 3'</u>. *All* BP *of degrees* $n \le 20$ *are irreducible, except, perhaps, for* $n = 15$.

<u>Proof.</u> The statement is obvious for $n = 1$ and $n = 2$. For $3 \le n \le 20$, $n \ne 15$, all odd integers are prime powers and all even integers are of the form $p \pm 1$. As for the irreducibility of $z_{15}(x)$, see Section 5.

(iii) In the more general case $n = qp-1$, $q < p/2$, one has $n = (q-1)p+(p-1)$,

$\sigma(n) = p+q-2$, $2n = (2q-1)p+(p-2)$, $\sigma(2n) = 2q+p-3$, $e(c_n) = (2n-\sigma(2n)-(n-\sigma(n))/(p-1)=q$, and $(n,e(c_n)) = (pq-1,q) = 1$. For $1 \leq m \leq n-1$, $m = rp+s$, $0 \leq r < q$, one has $\sigma(m) = r+s$. Also, $n+m = (r+q)p+(s-1)$, with $\sigma(m+n) = r+q+s-1$, if $s \neq 0$; $n+m = (r+q-1)p+(p-1)$, with $\sigma(m+n) = r+q+p-2$, if $s = 0$. Finally, $n-m = (q-r-1)p+(p-s-1)$ with $\sigma(n-m) = p+q-r-s-2$. By Lemma 2, $e(c_n) = r+1$ if $s \neq 0$, $e(c_n) = r$ if $s = 0$. Consequently, the line of slope $1/p$ through the origin runs through all spots $[rp,r]$, corresponding to the values $m = pr$ (with $s = 0$), while all other spots including $P_n = [n,e_n] = [(q-1)p+(p-1),q]$, are above that line. It follows that the N.p. of $z_n(x)$ consists of two sides, one from the origin to $P_{(q-1)p} = [(q-1)p,q-1]$, and the other from $P_{(q-1)p}$ to P_n. The first side has length $h_1 = (q-1)p$, and $v_1 = q-1$, $t_1 = (h_1,v_1) = q-1$, while the second has $h_2 = p-1$, $v_2 = 1$, $t_2 = (h_2,v_2) = 1$ and slope $\frac{1}{p-1} > \frac{1}{p}$, as required by the convexity (downward) of the N.p. By Theorem A, all factors of $z_n(x)$ have degrees of the form $d = \mu p + \nu(p-1) = (\mu+\nu)p-\nu$, $\mu = 0,1,\ldots,q-1$; $\nu = 0,1$. With this the proof of the first and of the last statement of Theorem 1 (b) is complete.

(iv) The treatment of the general case $n = qp+1$, $q < p/2$, is entirely analogous. One finds, successively, that $\sigma(n) = q+1$, $2n = 2qp+2$, $\sigma(2n) = 2q+2$, and $e(c_n) = q$. For $m = rp+s$ $(1 \leq m \leq n-1)$, one obtains that $0 \leq r \leq q$, with $r = q$ only if $s = 0$, $\sigma(m) = r+s$. Next, $n+m = (q+r)p+(s+1)$, with $\sigma(m+n) = q+r+s+1$, if $0 \leq s \leq p-2$;

$$n+m = (q+r+1)p, \qquad \text{with } \sigma(m+n) = q+r+1, \qquad \text{if } s = p-1.$$

Furthermore, $n-m = (q-r)p+1(1-s)$, with $\sigma(n-m) = q-r+1-s$, if $s = 0,1$;

$$n-m = (q-r-1)p+(p+1-s), \text{ with } \sigma(n-m) = q-r+p-s, \text{ if } 2 \leq s \leq p-1.$$

This leads to $$e(c_m) = \begin{cases} r & \text{if } s = 0, \text{ or } 1, \\ r+1 & \text{if } 2 \leq s \leq p-2, \\ r+2 & \text{if } s = p-1. \end{cases}$$

In particular, for $m = 1$ (i.e., $r = 0$, $s = 1$), $e(c_1) = 0$. As under (iii) we observe that the straight line through $P_1 = [1,0]$, with slope $1/p$ contains all spots with $m = rp+1$, $e(c_m) = r$, including $n = qp+1$, while all other spots lie above that line. Consequently, the N.p. of $z_n(x)$ consists now of two sides, one horizontal, from $P_0 = [0,0]$ to $P_1 = [1,0]$, of length one, and the second from P_1 to $P_n = [n,q]$ of length $h = n-1 = qp$, $v = q$, of slope $1/p$, and $t = (h,v) = (qp,q) = q$. It follows, by Theorem A, that any factor of $z_n(x)$ has a degree $d = \mu p+\nu$, with $\mu = 0,1,\ldots,q$ and $\nu = 0,1$. With this the first and the last statement of Theorem 1 (c) are proved.

(v) Let $n = qp$, with $q < p/2$; then $\sigma(n) = q$, $2n = 2qp$, $\sigma(2n) = 2q$, $e(c_n) = \frac{n-q}{p-1} = q$. Also, for $m = rp+s$, with $0 \leq r \leq q-1$, $0 \leq s \leq p-1$, $r+s \geq 1$, one has $\sigma(m) = r+s$ and, as before, computes $\sigma(n+m) = r+q+s$, $\sigma(n-m) = q-r$ if $s = 0$, $\sigma(n-m) = q-r-s-1$ if $1 \leq s \leq p-1$. By Lemma 2, $e(c_m) = r+1$ if $s \neq 0$, $e(c_m) = r$ if $s = 0$. The line of slope $1/p$ through the origin contains all the spots $[rp,r]$, including $P_n = [qp,q]$, while all other spots $[rp+s, r+1]$, $s \geq 1$, lie above that line. The N.p. consists of a single side, from $P_0 = [0,0]$ to $P_n = [n,q]$, with $h = qp$, $v = q$, and $t = (qp,q) = q$. By Theorem A, all factors of $z_n(x)$ have degrees of the form $d = \mu p$ ($\mu = 1,2,\ldots,q$). This finishes the proof of the first statement of Theorem 1 (d).

(vi) The next type of n to be considered is $n = p^k-1$. Now $n = (p-1)(p^{k-1}+\ldots+p-1)$, so that $\sigma(n) = k(p-1)$. Also, $2n = 2p^k-2 = p^k + \sum_{j=1}^{k-1}(p-1)p^j+(p-2)$ and $\sigma(2n) = \sigma(n)$. By Lemma 2, $e(c_n) = \frac{n}{p-1} = \sum_{j=0}^{k-1} p^j$. For $m = \sum_{j=0}^{k-1} a_jp^j$, $\sigma(m) = \sum_{j=0}^{k-1} a_j$. Also, $n-m = \sum_{j=0}^{k-1}(p-1-a_j)p^j$, with $\sigma(n-m) = k(p-1)-\sigma(m)$. Finally, $n+m = p^k + \sum_{j=0}^{k-1} a_jp^j-1$. If $a_j = 0$ for $j = 0,1,\ldots,r-1$, while $a_r \neq 0$, then $n+m = p^k + \sum_{j=r+1}^{k-1} a_jp^j + (a_r-1)p^r + (p-1)\sum_{j=0}^{r-1} p^j$ and $\sigma(n+m) = \sum_{j=r}^{k-1} a_j+r(p-1) = \sigma(m)+r(p-1)$. By Lemma 2, $e(c_m) = \frac{1}{p-1}\{m+\sigma(m)+k(p-1)-\sigma(m)-\sigma(m)-r(p-1)\} = \frac{1}{p-1}\{m-\sigma(m)\} + k-r$. We shall verify that

$$(5)\qquad \frac{me(c_n)}{n} = \frac{m}{p-1} \leq e(c_m),$$

with equality if, and only if for all $j \geq r$, $a_j = p-1$. Written explicitly, (5) reads $m-\sigma(m) + (p-1)(k-r) \geq m$, or

$$(6)\qquad (p-1)(k-r) \geq \sigma(m) = \sum_{j=r}^{k-1} a_j.$$

For given r, the right hand side is maximized by the choice of $a_j = p-1$. In that case, and that case only, (6) becomes an equality. For any other choice (6) is a strict inequality. This shows that all the spots that correspond to $m = (p-1)\sum_{j=r}^{k-1} p^j$ ($r = 0,1,\ldots,k-2$) lie on the straight line $y = (e(c_n)/n)x = x/(p-1)$, of slope $1/(p-1)$ through the origin, while all other spots are above that line.

The N.p. of $z_n(x)$ consists, therefore of a single side with $h = n = p^k-1$, $v = e(c_n) = \frac{n}{p-1} = \frac{p^k-1}{p-1}$, $t = (h,v) = \frac{p^k-1}{p-1}$, and $s = n/t = p-1$. According to Theorem A, all factors of $z_n(x)$ have degrees of the form $\mu(p-1)$ ($\mu = 1,2,\ldots,t$). This proves the first statement of Theorem 1 (e).

REMARK. A slightly stronger form of Theorem A can be obtained, by the remark that any splitting of f(x) into irreducible factors can be effected by splitting f(x) first into two (not necessarily irreducible) factors, then splitting any factor that is not already irreducible into two factors, etc. At each stage the sum of the degrees of the factors equals the degree of the polynomial being split. By keeping track of this equality one can obtain the mentioned stronger version of Theorem A. Before stating it, it is convenient to define a new term. Whenever there are spots on a side of a N.p., these divide the side into collinear segments, that we shall call (following Wahab [60]), the *elements* of the N.p. If $b_1,b_2,\ldots,b_g$ are the lengths (i.e., the horizontal projections) of the different elements, then we have the following version of Theorem A, which in a somewhat different formulation appears to be due to Wahab.

THEOREM A'. (Wahab [60]). *Let* $f(x) = f_1(x)f_2(x)\ldots f_\ell(x)$; *then* d_s, *the degree of* $f_s(x)$ *is of the form* $d_s = \sum_{j=1}^{g} \delta_{sj}b_j$, *with* $\delta_{sr} = 0,1$ *and* $\delta_{sr}\delta_{tr} = 0$ *if* $s \neq t$.

For a complete proof of Theorem A', see [60].

If we use this version of Theorem A, we have some added information concerning the possible degrees of the factors of $z_n(x)$, $n = p^k-1$. Indeed we already know that the spots on the single side of the N.p. of $z_n(x)$ have the abscissae $(m-1)\sum_{j=r}^{k-1} p^j$ $(r = 0,1,\ldots,k-2)$. It follows that the lengths of the elements are $b_1 = p^k-p^{k-1}$, $b_2 = p^{k-1}-p^{k-2},\ldots,b_k = p-1$. By Theorem A', any factor of $z_n(x)$ has a degree $d_s = \sum_{j=1}^{k} \delta_{sj}(p^{k-j+1}-p^{k-j})$ ($\delta_{sj} = 0$ or 1). This result, not needed in the proof of Theorem 1, is mentioned here, because it seems that this kind of closer analysis may lead eventually to a proof of Conjecture 1.

(vii) By (3), $c_m = \frac{(n+m)!}{m!(n-m)!} = \frac{(n-m+1)(n+m)}{m} c_{m-1}$ and $c_{m+1} = \frac{(n-m)(n+m+1)}{(m+1)} c_m$.

It follows that (without the earlier restriction $q < p/2$), $e(c_m) = e(c_{m-1})$, unless p divides at least one of the factors $n-m+1$, $n-m$, or m. For $p|n$ this can happen only for $m \equiv 0$, or 1 (mod p). This means that the spots arrange themselves in horizontal rows of equal $e(c_m)$'s with steps up, or down, only for $m \equiv 0,1 \pmod p$. From the fact that the N.p. of any $z_n(x)$ starts at $P_0 = [0,0]$ and the downward convexity of the N.p. it follows that if $[m_1, e(c_{m_1})]$ is a spot on a side of the N.p., with $e(c_{m_1}) > 0$, all other spots with $m_2 > m_1$ have $e(c_{m_2}) > e(c_{m_1})$. For $e(c_{m_1}) = 0$ no such restriction exists and we have spots at all integers m with $e(c_m) = 0$. It furthermore follows that a spot with $e = e(c_{m_2})$ can be a vertex, or indeed even a spot on a side of a N.p. only if m_2 is the largest value of m for which $e(c_m) = e(c_{m_2})$ (see Fig. 2).

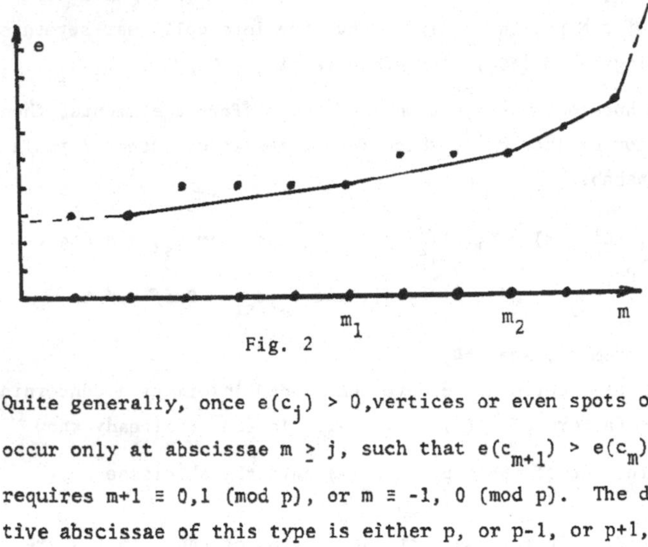

Fig. 2

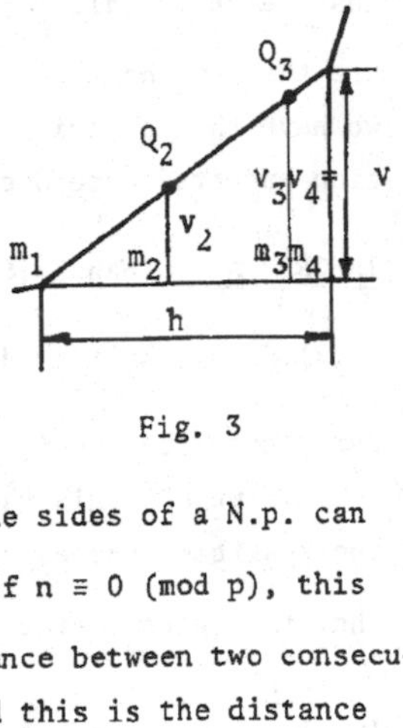

Fig. 3

Quite generally, once $e(c_j) > 0$,vertices or even spots on the sides of a N.p. can occur only at abscissae $m \geq j$, such that $e(c_{m+1}) > e(c_m)$. If $n \equiv 0 \pmod p$, this requires $m+1 \equiv 0,1 \pmod p$, or $m \equiv -1, 0 \pmod p$. The distance between two consecutive abscissae of this type is either p, or p-1, or p+1, and this is the distance between consecutive spots on a side of the N.p. This does not necessarily mean that the length of the complete side of the N.p. has one of these values. It is quite possible to have a side continue in the same direction and hit another spot after another interval of length p-1, p, or p+1. The side of the N.p. cannot continue with a *smaller* slope than on a preceding interval, as that would violate the convexith downwards, but it may continue with the same slope.

After the side has traversed a number of spots, it either reaches $P_n = [n,e(c_n)]$ and the N.p. terminates, or else the polygonal line bends again upwards, thus starting a new side of the N.p.

We may verify that the spots on any side lead to elements whose lengths satisfy the requirement of Theorem A namely to be of the form μs, $\mu \leq t = (h,v)$. To do this we observe that along a given side the spots that occur as interior points, say Q_2, Q_3 in Fig. 3 are lattice points (they have integral coordinates). No such points can appear, if $(h,v) = 1$. Indeed, $v_j = (v/h)(m_j-m_1)$ with $h > m_j-m_1$ for m_j not the last abscissa so that v_j cannot be an integer, unless v/h is a reducible fraction, i.e. $(h,v) = t > 1$. If $v = tw$, $h = ts$, $(s,w) = 1$, then $v_j = (w/s)(m_j-m_1)$ with $s|(m_j-m_1)$, so that $m_j-m_1 = \mu_j s \leq h = ts$, $\mu_j \leq t$, $v_j = w\mu_j$, and finally $m_j-m_1 = (v_j/w)s = \mu_j s$, $\mu_j \leq t$, in agreement with Theorem A. From the point of view of Theorem A', m_2-m_1, m_3-m_2 are the lengths of the elements of the N.p. It follows that the length of an element (hence, that of the permitted degrees of a factor) cannot be less than $p-1$.

In case $n \equiv -1 \pmod p$, the factors $n-m+1$, $n+m$, and m are congruent mod p to $-m$, $m-1$, and m, respectively. Hence, $e(c_{m+1}) > e(c_m)$ only for $m \equiv 0$, or $1 \pmod p$ and the N.p. can have spots on a side of the N.p., or vertices, only for $m \equiv -1$, or $0 \pmod p$. From here on the proof proceeds as before and its details may be suppressed. The proof of Theorem 1 (f) is complete.

5. A given integer n may belong to several of the types considered in Theorem 1. So, e.g., for $n = 9$, we may write $n = p^m$ ($p = 3$, $m = 2$), or $n = qp-1$ ($q = 2$, $p = 5$). The prime with respect to which one considers the N.p. will of course be different in each representation. In general, after selection of a definite prime p, Theorem A indicates factorizations that may be possible for a given n.

So, e.g., for $z_{15}(x)$, the N.p. w.r. to $p = 13$ has two sides; one of the sides, of length $h_1 = 2$, has a spot in the middle, the other one of length $h_2 = 13$ contains no spots. The N.p. contains 3 elements and, by Theorem A (or A'), any polynomial factor of $z_{15}(x)$ may only have degrees obtained as sums of the integers 1,1, and 13. We state such a result succintly in the form (15) = (1)(1)(13) and call it a *scheme of factorization*. The N.p. of the same $z_{15}(x)$ but with respect to $p = 3$ leads to (15) = (9)(4)(2). These two results, while different, are not contradictory. They can be reconciled in two ways. Either trivially, as (1+1+13) = (15) = (9+4+2) if $z_{15}(x)$ is irreducible and none of the potentially allowed factorizations actually exists, or else by observing that both schemes allow the non-trivial factorization (1+1)(13) = (2)(13) = (2)(9+4).

Two such factorization schemes, with a common, non-trivial factorization, will be called *compatible*. Otherwise, we call them incompatible. It is clear that if one can exhibit two incompatible factorization schemes, one thereby will have proved the irreducibility of the polynomial considered. In the case of $z_{15}(x)$, the N.p. w.r. to $p = 5$ leads to the scheme (15) = (5)(4)(6). Also this is compatible with (15) = (9)(4)(2), because (5+4)(6) = (9)(6) = (9)(4+2). It is, however, incompatible with the scheme (15) = (1)(1)(13) obtained from the N.p. w.r. to $p = 13$ and we conclude that $z_{15}(x)$ is irreducible.

There should be no need now for a more formal proof of the

<u>PROPOSITION 1.</u> *If the N.p. of* $z_n(x)$ *with respect to two different primes lead to incompatible schemes of factorization, then* $z_n(x)$ *is irreducible.*

6. In this section we shall prove Theorem 4.

This will be needed in the proof of Theorem 2 and only by use of Theorem 2 are we going to complete the proof of Theorem 1.

We already know that $z_n(x)$ is irreducible if $n \leq 20$. This follows from Theorem 3' and the irreducibility of $z_{15}(x)$ proved in Section 5. Also, $z_n(x)$ is irreducible if $n = p$. If n is not a prime, let $p_1 < n < p_2$, with p_1 and p_2 consecutive primes and set $n = p_1+k_1 = p_2-k_2-1$. Then $k_1 = n-p_1 < p_2-p_1$ and $k_2 = p_2-n-1 < p_2-p_1-1$ or, $k_2 \leq p_2-p_1-2$. From Theorem C we know that $p_2-p_1 \leq p_1/8$ for $p_1 \geq 53$. For $11 \leq p_1 < 53$, we verify that $p_2-p_1 < \frac{p_1}{2}$ (the largest value is attained for $p_1 = 23$, with $29-23 = 6 = (6/23)p_1$). This inequality holds by Theorem C for all $p_1 \geq 11$, which is more than sufficient because we are concerned only with $n \geq 21$ ($p_1 \geq 19$). We note, in particular, that $p_1+2k_1 < p_1+2(p_2-p_1) < 2p_1$. This insures that the N.p. of $z_n(x)$ w.r. to p_1 has the first k_1 coefficients not divisible by p_1. Indeed,

$$c_1 = \frac{n(n+1)}{1!},\ c_2 = \frac{(n-1)n(n+1)(n+2)}{2!},\ \ldots,\ c_{k_1} = \frac{(n-k_1+1)\ldots(n+k_1)}{k_1!} =$$

$$\frac{(p_1+1)\ldots(p_1+2k)}{k_1!} \quad \text{and } p_1 < (p_1+1),\ (p_1+2k) < 2p_1.$$

We claim that all other coefficients are divisible by p_1 to exactly the first power. For $k_1 < m < p_1$, the factor p_1 which now occurs in the numerator is not cancelled by any factor of the denominator. For $m = p_1$,

$$c_{p_1} = \frac{(k_1+1)(k_1+2)\ldots p_1 \ldots (2p_1)}{p_1!} \quad \text{and } p_1 | c_{p_1},\ p_1^2 \nmid c_{p_1} \quad \text{still holds.}$$

None of the factors by which c_{p_1} has to be multiplied in order to obtain the coefficients c_m, $p_1 < m \leq n$, contain multiples of p_1, either in the numerator, or the denominator, and the claim is proved. We may verify it, in particular, for $m = n$,

$c_n = (n+1)\ldots(2n) = (p_1+k_1+1)\ldots(2p_1)\ldots(2p_1+2k_1)$ with $p_1+k_1+1 > p_1$ and, $2p_1+2k_1 < 3p_1$. It follows that the N.p. of $z_n(x)$ w.r. to p_1 consists of 2 sides: the first one of length k_1, on the real axis, with spots at all the integers and the second one from $P_{k_1} = [k_1,0]$ to $P_n = [n,1]$, with no spots on it and of length $h_2 = n-k_1 = p_1$. By Theorem A this leads to the factorization scheme $(n) = \underbrace{(1)(1)\ldots(1)}_{k_1 \text{ times}}(p_1)$ and it follows that $z_n(x)$ contains an irreducible factor of degree at least equal to p_1.

We study in exactly the same way the N.p. of $z_n(x)$ w.r. to p_2. In particular,

$$c_{k_2} = \frac{(n-k_2+1)\ldots n(n+1)\ldots(n+k_2)}{k_2!} = \frac{(p_2-2k_2-1)\ldots(p_2-k_2-1)(p_2-k_2)\ldots(p_2-1)}{k_2!}$$

and all coefficients c_m with $0 \le m \le k_2$ are not divisible by p_2, while

$$c_{k_2+1} = \frac{(p_2-2k_2)\ldots(p_2-1)p_2}{(k_2+1)!}$$ and all successive ones contain p_2 to exactly the first power. Indeed, $n < p_2$, so that no factor of the denominator can cancel the factor p_2 in the numerator. In particular, $c_n = (n+1)\ldots(2n) = (p_2-k_2)\ldots p_2\ldots(2p_2-2k_2-2)$ and the last factor is less than $2p_2$. It follows as above that $z_n(x)$ contains an irreducible factor of degree at least equal to $n-k_2$. If we combine this result with the preceding one and recall the notation $k = k(n) = \min(k_1,k_2)$, we can state

THEOREM 4. *For every* n, *the* BP *of degree* n *contains an irreducible factor of degree at least equal to* n-k(n) *and can contain no factor of degree* d *with* $k(n) < d < n-k(n)$

7. Proof of Theorem 2. As already seen in Section 2, for every $\varepsilon > 0$ and $x \ge N(\varepsilon)$, there is at least one prime p in the interval $x < p \le (1+\varepsilon)x$. In particular, it follows with $x = p_1$, that there is at least one prime p_2 such that $p_1 < p_2 \le (1+\varepsilon)p_1$. Let $p_1 < n < p_2$ with p_1, p_2 consecutive primes; we want to estimate the largest possible value of k(n). In $k_1 = n-p_1$ and $k_2 = p_2-n-1$, set $n = r_1p_1$, $p_2 = nr_2$, so that $k_1 = n(1-1/r_1)$, $k_2 = n(r_2-1-1/n)$. Now, k(n), the smaller one of k_1,k_2, takes its largest possible value, if n happens to fall between p_1 and p_2 in such a way as to make $k_1 = k_2$. This means, $1-r_1^{-1} = r_2-1-n^{-1}$. Also, $np_2 = nr_1r_2p_1$, whence $r_1r_2 = p_2/p_1 \le 1+\varepsilon$. The most unfavorable situation (large k) occurs when the

separation between p_1 and p_2 is largest, i.e. for $p_2/p_1 = 1+\varepsilon$. By solving the system of 2 equations obtained for r_1, r_2, we find $r_1 = \frac{2+\varepsilon}{2+1/n}$, $r_2 = \frac{(2+1/n)(1+\varepsilon)}{2+\varepsilon}$. From this we obtain $k_1 = n-p_1 \le n(1-r_1^{-1}) = \frac{n\varepsilon-1}{2+\varepsilon}$ and verify that also $k_2 = p_2-n-1 \le n(r_2-1-n^{-1}) = \frac{n\varepsilon-1}{2+\varepsilon}$. Consequently, $k = k(n) \le \frac{n\varepsilon-1}{2+\varepsilon} < n\frac{\varepsilon}{2+\varepsilon}$. The fraction k/n can be made arbitrarily small by taking ε sufficiently small. We may indeed take $\varepsilon > 0$ arbitrarily small, provided we take n sufficiently large, $n \ge N(\varepsilon)$. By Theorem 4, the degree of an irreducible factor of $z_n(x)$ exceeds $n-k(n) \ge n(1-\frac{\varepsilon}{2+\varepsilon}) = A_n n$, say, with $A_n = 1-\frac{\varepsilon}{2+\varepsilon} = \frac{1}{1+\varepsilon/2}$.

For $n \to \infty$, one has $\varepsilon \to 0$, hence $\lim_{n\to\infty} A_n = 1$ and this proves the first statement of Theorem 2.

We now recall that, by Theorem C, we may take $\varepsilon = 1/8$ for $n \ge p_1 \ge 53$. Hence, for $n \ge 53$, $\frac{k(n)}{n} < \frac{1/8}{2+1/8} = \frac{1}{17}$ and $A_n \ge \frac{1}{1+1/16} = \frac{16}{17}$; this proves the validity of Theorem 2.

8. Proof of Theorem 1. Part II. It is now easy to complete the proof of Theorem 1. Let us assume first that $n \ge 53$. We shall show that under the assumptions of the theorem the N.p. w.r. to p (of Theorem 1) and the N.p. w.r. to p_1 or p_2 (largest prime less than n, or least prime larger than n) lead to incompatible schemes of factorization, so that the irreducibility follows from Proposition 1.

If we assume that $a_n(x)$ is not irreducible and $n = qp-1$, $q < p/2$, as assumed in Theorem 1 (b), then the lowest possible degree of an irreducible factor, satisfies, as seen, $d \ge rp-1 \ge p-1$. From Theorem 4 it follows that $d \le k(n)$, so that $p-1 \le k(n)$. Under the assumption $p-1 > k(n)$, this is not possible, so that $z_n(x)$ is irreducible. Similarly, under the conditions of Theorem 1 (c), if $z_n(x)$ is not irreducible, the minimal degree d of one of its factors satisfies either $d = 1$ or $d \ge p$. By Theorem 4, $d \le k(n)$, so that $d \ge p$ implies $p \le k(n)$. If we know, however, that $p > k(n)$, the last inequality is impossible, so that $z_n(x)$ must be either irreducible or else has a linear factor with a real (in fact, integral) zero. If q is odd, then $qp+1$ is even and (see Chapter 10) the last alternative cannot occur.

Exactly the same reasoning, applied to the conditions of Theorem 1 (d) and (e), with $p > k(n)$ or $p-1 > k(n)$, respectively,proves the irreducibility in those cases.

If, instead of $p > k(n)$, or $p > k(n) + 1$ we know instead that $q \le 17$, then, say, in Theorem 1 (d) we obtain $p = n/q \ge n/17$. As proved in Section 7, $k(n) < n/17$, so that $p > k(n)$. If $z_n(x)$ is not irreducible, then all factors have degrees equal to μp, $\mu p \ge p > k(n)$, and this contradicts Theorem 4 and proves the irreducibility of

$z_n(x)$. It is clear that, while not stated in the Theorem, the condition $q \leq 17$ can be substituted for $p-1 > k(n)$ also in Theorem 1 (b) and (c). With this, Theorems 1 and 2 are completely proved, at least for $n \leq 20$ and $n \geq 53$.

9. It still remains to be shown that $z_n(x)$ is irreducible for $21 \leq n \leq 53$. In the proof we may use all previous results, except, of course, those statements of Theorems 1 and 2, such as those referring to $q \leq 17$, that were obtained under the assumption $n \geq 53$.

From Theorem 1 (a,b,c), we know that for all $n = p^m$ and $n = p \pm 1$, the $z_n(x)$ are irreducible (without the restriction $n \geq 53$). After we eliminate these n, we still remain with eight values. For each of them it is easy to find $k = k(n)$ by inspection. So, e.g., for $n = 33$, $p_1 = 31$, $p_2 = 37$, $k_1 = 2$, $k_2 = 3$ and $k(33) = 2$. For five of these 8 values of n we may use Theorem 1(d) as follows: $21 = 3.7$, $k = 1$; $33 = 3.11$, $k = 2$; $34 = 2.17$, $k = 2$; $39 = 3.13$, $k = 1$; $51 = 3.17$, $k = 1$. In each case any factor would have to have a degree d that is a multiple of the larger of the two prime factors of n, and at the same time have $d \leq k$. For instance, if $n = 33$, $d = \mu.11 \geq 11$ and $d \leq 2$. This is not possible and hence, $z_{33}(x)$ is irreducible. The last remaining three values of n are:

$35 = 2.17+1 = 5.7$, with factors of degrees $17r$, or $17r+1$ only, $k = 1$, and also with the factorization $(35) = (10)(25) \pmod 5$;

$45 = 4.11+1 = 3^2.5$, with factors of degrees $11r$, or $11r+1$ only, $k = 1$, and also with the factorization $(45) = (25)(4)(16) \pmod 5$;

$50 = 3.7-1$, with factors of degrees $7r$, or $7r-1$ only, and $k = 2$.

By Theorem A and Theorem 4, all corresponding polynomials $z_n(x)$ are irreducible. With this Theorem 1 and Theorem 2 are completely proved.

10. <u>Proof of Theorem 3.</u> We already proved a weak form of Theorem 3, namely

<u>THEOREM 3"</u>. *All* BP *of degree* n, *with* $n \leq 53$ *are irreducible.*

In order to complete the proof of Theorem 3, we proceed as before. We first suppress from the list of all integers n, $53 < n \leq 400$, all primes, odd prime powers, and integers of the form $p \pm 1$. For the remaining integers n, one determines $k(n)$, by looking for the primes closest to n. Next, one tries to fit each n into one of the patterns of Theorem 1, such as $n = qp+\delta$, $\delta = -1, 0, +1$, $q < p/2$ or $n = p^m-1$. This will impose upon the smallest degree d of a factor of $z_n(x)$ the condition $d \geq p$, $d \geq p-1$ or $d = 1$. On the other hand, $d \leq k(n)$ and, if $d \neq 1$ and $k(n) < p-1$, the resulting contradiction is proof of the irreducibility of the corresponding $z_n(x)$ If $d = 1$ and n is even no linear factor may split off, with the same result.

After the odd prime powers and $n = p \pm 1$ have been eliminated, there remain 61 integers n with $53 < n \leq 200$. These are listed in the Table 1, each with at least one representation $n = qp+\delta$ or $n = p^m-1$. In all cases $p \geq 11$ and k(n), which is not listed, never exceeds 6.

Table 1

55 = 5.11	115 = 5.23	154 = 5.31-1
56 = 3.19-1	116 = 4.29	155 = 5.31
57 = 3.19	117 = 2.59-1	159 = 3.53
63 = 2.31+1 = 3^2.7	118 = 2.59	160 = 7.23-1
64 = 5.13-1	119 = 7.17	161 = 7.23
65 = 5.13	120 = 11^2-1	165 = 2.83-1
69 = 3.23	122 = 2.61	170 = 9.19-1
75 = 4.19-1	123 = 3.41	171 = 9.19
76 = 4.19	124 = 4.31	175 = 6.29+1 = 7.5^2
77 = 6.13-1	129 = 3.43	176 = 3.59-1
85 = 5.17	133 = 7.19	177 = 3.59
86 = 2.43	134 = 2.67	183 = 3.61
87 = 3.29	135 = 8.17-1	184 = 5.37-1
91 = 4.23-1	141 = 3.47	185 = 5.37
92 = 4.23	142 = 2.71	186 = 6.31
93 = 3.31	143 = 2.71+1 = 11.13	187 = 4.47-1 = 11.17
94 = 2.47	144 = 5.29-1	188 = 4.47
95 = 5.19	145 = 5.29	189 = 10.19-1
99 = ?	146 = 2.73	195 = 2.97+1 = 15.13
105 = 2.53-1	147 = 4.37-1	
111 = 3.37	153 = 8.19+1 = 9.17	

From Table 1 it appears that all integers up to 200 fit readily into one of the patterns of Theorem 1, except for n = 99. The polynomial $z_{99}(x)$, however, is irreducible by Theorem 1 (f) and Theorem 4. Indeed, $p_m = 11$ is the largest prime divisor of 99, so that $z_{99}(x)$ cannot have factors of degrees less than 10. On the other hand, $k(99) = k_2 = 101-99-1 = 1$ and $z_{99}(x)$, if not irreducible, can split at most into a linear factor and an irreducible factor of degree 98, whence the irreducibility follows. Also for n = 63, 143, 175, and 195, all odd and of type $n = qp+1$ $(q < p/2)$ one must still show that they do not contain linear factors. For both n = 63 and n = 175, the largest prime factor is $p_m = 7$ and by Theorem 1 (f) no factors of degrees less than 6 exists. For n = 143 and n = 195, $p_m = 13$ and no factors of degrees less than 12 exist.

While not needed for the purpose on hand, it is instructive to see to what factorization schemes one is led in these 5 cases, by the N.p. w.r. to the corresponding largest prime factor p_m. One obtains (99) = (11)(10)(78), (63) = (7)(7)(49), (143) = (13)(12)(118), (175) = (7)(7)(49)(49)(49)(6)(8) and (195) = (13)(13)(169).

One may remark that, while n = 63 and n = 195 lead to $d \geq p_m$, for n = 99, 143, and 175, one obtains nothing better than $d \geq p_m-1$, so that the statement of Theorem 1 (f) cannot be improved without some added restrictions.

The verification that also all $z_n(x)$ with $200 < n \leq 400$ are irreducible is equally easy - and dull. For $201 \leq n \leq 300$, all integers fit one of the patterns of Theorem 1, except for 209. Here $p_m = 19$, $k(209) = 1$, so that $z_{209}(x)$ is irreducible by Theorem 1 (f) and Theorem 4. In $301 \leq n \leq 400$ we find four values of n that have to be considered separately, namely n = 323 ($p_m = 19$, k = 6), 324 ($324+1 = 5^2.13$, $p_m = 13$, k = 6), 351 ($p_m = 13$, k = 1) and 391 ($p_m = 23$, k = 2). Three of them are irreducible by Theorem 1 (f) with n = qp and Theorem 4, while for n = 324 one uses Theorem 1 (f) with n = qp-1 and Theorem 4.

In spite of the ease with which we are able to prove the irreducibility of any single polynomial $z_n(x)$, a general method still eludes us. Such examples as n = 99 and n = 209, etc., show that we cannot take it for granted that we shall be able to fit all integers n, into any finite set of patterns, for which the irreducibility can be proved. By working out a large number of N.p., one is led to make certain conjectures.

There also is some hope that the method used in the proof of Theorem 1 (e) and 1 (f) could be perfected, in order to prove

<u>CONJECTURE 2.</u> If $p' \neq p''$ and $p'p''|n$, then the N.p. of $z_n(x)$ w.r. to p' and to p'' lead to incompatible schemes of factorization.

It is clear that the proof of Conjecture 2 would go a long way towards the proof of Conjecture 1. In fact, the only integers not covered would be those of the form $n = 2^m p^k$ and which also cannot be represented as $n = qp''' \pm 1$, with appropriately small q.

CHAPTER 12

THE GALOIS GROUP OF B.P.

1. The Galois group of the BP of degree n will be denoted by G_n, the symmetric group on n symbols by S_n and the corresponding alternating group by A_n.

A group is said to be of degree n, if it is isomorphic to a permutation group on n symbols. The terms "transitive", "primitive", etc., have their usual meaning from group theory as defined, e.g., in [9]. The term irreducible will mean irreducible over the rationals, as in Chapter 11.

The material of this chapter is essentially contained in [53, Section 5] and in [54], but several proofs have been reworked.

The main result is

THEOREM 1. *For every* n, *if* G_n *is transitive, then* $G_n = S_n$.

This theorem can be rephrased. Indeed, the Galois group of a polynomial is transitive, if and only if the polynomial is irreducible. Hence, we may state

THEOREM 1'. *If* $y_n(z)$ *is irreducible, then* $G_n = S_n$.

We saw in Chapter 11, that a large class of BP consists in fact of ireducible polynomials and we formulated there Conjecture 1, according to which all BP are irreducible. This leads us to state

THEOREM 1''. *Conjecture* 1 *of Chapter* 11 *implies that, for every* n, $G_n = S_n$.

2. Theorem 1' is equivalent to Theorem 1 and implies Theorem 1''. Hence, it is sufficient to prove Theorem 1'. For that we shall need several known results. As in Chapter 11, we shall have to content ourselves with the statements only and refer the reader to easily available proofs in the literature.

THEOREM A. (I. Schur [52], Theorem V). *Let*

$$f(x) = \sum_{j=0}^{n} a_j x^{n-j}, \quad a_j \text{ rational integers,}$$

have the discriminant D *and be irreducible, and let* G_n *be its Galois group. Let us assume that there exists a prime* p, *which satisfies the following conditions:*

$$p^m \mid D \text{ with } m \geq n,\ p \mid a_n,\ p^2 \nmid a_n.$$

Assume also that if $f(x) \equiv x^k g(x) \pmod p$ *and* d *is the discriminant of* g(x), *then* $p \nmid d$. *Any such prime is a divisor of the order of* G_n.

If, furthermore, $n/2 < p < n-2$, *then* $G_n = A_n$ *in case* D *is a perfect square,* $G_n = S_n$, *otherwise.*

THEOREM B. (Dedekind; see [52]). *Let us suppose that the polynomial* $f(x)$ *splits modulo* p *into a product of* r *irreducible polynomials* $f_i(x)$ $(i = 1,2,\dots,r)$, *incongruent in pairs modulo* p. *Then the Galois group of* $f(x)$ *contains at least one permutation of* r *cycles, such that each cycle corresponds to one of the factors* $f_i(x)$, *and the order of each cycle is equal to the degree of the polynomial to which it corresponds.*

THEOREM C. (Burnside; see [9],XI). (1) *If* n (> 3) *and* $2n+1$ *are both primes, then any triply transitive group of degree* $2n+3$ *contains the alternating group* (§ 165, page 214, Exercise).
(2) *For* $n = 4$, *there are no primitive groups, except* A_4 *and* S_4 (§ 166, page 214, (ii)).
(3) *Every triply transitive group of degree seven contains* A_7 (§ 166, pages 216-218, (v); see also [52(a)], p. 449, or [52(b)], p. 197).
(4) *If* p^e *is the highest power of* p *contained in* $n!$ *and if* $p < 2n/3$, *then* p^{e-1} *is the highest power of* p *that divides the order of any primitive group of degree* n, *which does not contain* A_n (§ 160, pages 207-208).

THEOREM D. (Cauchy; see [30], p. 74). *If* p *divides the order of a group* G, *then* G *contains an element of order* p.

THEOREM E. (Jordan [39], Note C; see also [9], p. 214, Theorem I and [52](a), p.448; [52(b)], p. 196). *If* G *is a primitive transitive group of degree* n *and* $p > \frac{n}{2}$ *is a divisor of the order of* G, *then the degree of transitivity of* G *is at least* $n-p+1$.

3. The method of proof follows, in general outline that of Schur ([52]), and it is largely based on Theorem A. We also shall need the following

LEMMA 1. *For* $(n+1)/2 \le p < n$, $\theta_n(x) \equiv x^p\theta_{n-p}(x) \pmod p$.

Proof. Trivially $n \equiv (n-p) \pmod p$; consequently, for $m < n-p+1$, one has $m < n - (n+1)/2+1 = (n+1)/2 \le p$, so that

$$a_m^{(n)} = \frac{(n-m+1)\dots n(n+1)\dots(n+m)}{2^m m!}$$

$$\equiv \frac{((n-p)-m+1)\dots(n-p)(n-p+1)\dots((n-p)+m)}{2^m m!} = a_m^{(n-p)} \pmod p.$$

On the other hand, for $n-p \le m < p$, $a_m^{(n)}$ contains the factor p in the numerator, but not in the denominator and $a_m^{(n)} \equiv 0 \pmod p$. For $p \le m \le n$, the denominator contains the factor p to exactly the first power (because $n \le 2p-1 < 2p$), while the numerator contains at least two multiples of p (p itself and 2p, because $n \ge p$,

$m \geq p$ imply $n+m \geq 2p$). It follows that

$$\theta_n(x) = \sum_{m=0}^{n} a_m^{(n)} x^{n-m} \equiv \sum_{m=0}^{n-p} a_m^{(n)} x^{n-m} \equiv \sum_{m=0}^{n-p} a_m^{(n-p)} x^{(n-p)-m+p}$$

$$\equiv x^p \sum_{m=0}^{n-p} a_m^{(n-p)} x^{(n-p)-m} \equiv x^p \theta_{n-p}(x) \pmod p.$$

We now verify that the conditions of Theorem A hold for $\theta_n(x)$. Specifically, we shall do the following:

(a) compute the discriminant D_n of $\theta_n(x)$;

(b) show that for $n \geq 14$, and also for $n = 10$, there exists a prime p such that $(2n-1)/3 < p < n-2$, and also $p^n | D_n$, $p | a_n^{(n)}$, $p^2 \nmid a_n^{(n)}$;

(c) show that $\theta_n(x) \equiv x^p \theta_{n-p}(x) \pmod p$ and $p \nmid D_{n-p}$;

(d) verify that D_n is not a perfect square.

Theorem A can be applied to those values of n, for which these conditions can be verified (in fact, the present condition (b) is more stringent than a corresponding condition of Theorem A) and we conclude that for those n, $G_n \supset A_n$. From (d) it follows, however, that $G_n \neq A_n$, so that indeed $G_n = S_n$.

If those conditions hold for some p that falls outside the range required by (b), then it still follows from Theorem A, that p divides the order of G_n. By Theorem D it then follows that G_n contains an element of order p. If also $p > n/2$, then the group G_n (transitive by assumption) is obviously primitive. Theorem E then insures a rather high degree of transitivity, which, in some cases, suffices to show that $G_n \supset A_n$, so that, by (d) above, $G_n = S_n$.

In the few remaining cases we shall succeed to reach that conclusion, by appealing to Theorem B.

4. (a) We recall (see, e.g. [59], § 28) that R(f,g), the resultant of two polynomials $f(x) = \sum_{i=0}^{n} c_i x^{n-i}$ of degree n and $g(x) = \sum_{j=0}^{m} d_j x^{m-j}$ of degree m, is a polynomial function of the coefficients c_i, d_j ($i = 1,2,\ldots,n$; $j = 1,2,\ldots,m$) that vanishes if and only if f and g have a common zero. In fact, $R(f,g) = c_0^m d_0^n \prod_{i=1}^{n} \prod_{j=1}^{m} (x_i - y_j)$, where the x_i range over the zeros of f(x) and the y_j over those of g(x). Clearly, $R(f,g) = c_0^m \prod_{i=1}^{n} g(x_i) = (-1)^{nm} d_0^n \prod_{j=1}^{m} f(y_i) = (-1)^{nm} R(g,f)$. We shall be concerned here with monic polynomials, so that $c_0 = d_0 = 1$, and nm will always be even, so

that these formulae simplify accordingly.

The discriminant D of a polynomial f(x) is, essentially, the resultant of f(x) and f'(x). More precisely,

(1) $$R(f,f') = (-1)^{n(n-1)/2} c_0 D.$$

It follows that $D = c_0^{n-2} \prod_{i=1}^{n} f'(x_i) = (-1)^{n(n-1)/2} c_0^{2n-2} \prod_{i \neq j} (x_i - x_j) =$ $c_0^{2n-2} \prod_{i<j} (x_i - x_j)^2$, where the x_i's and x_j's run through the zeros of f(x). The last expression is often taken as definition of D.

In order to compute D_n, the discriminant of $\theta_n(x)$, we first compute $R(\theta_{n-1}, \theta_n)$. By (3.5), $\theta_{n+1}(x_i) = x_i^2 \theta_{n-1}(x_i)$ for each zero x_i of $\theta_n(x)$. By taking the product over all these zeros and recalling also that θ_n is monic, we obtain

$$R(\theta_n, \theta_{n+1}) = \prod_{i=1}^{n} \theta_{n+1}(x_i) = (x_1 x_2 \ldots x_n)^2 \prod_{i=1}^{n} \theta_{n-1}(x_i) = (a_n^{(n)})^2 R(\theta_n, \theta_{n-1}).$$

Either n, or n-1 is even; hence $R(\theta_n, \theta_{n-1}) = R(\theta_{n-1}, \theta_n)$ and we obtain by an obvious inductive argument, $R(\theta_n, \theta_{n+1}) = \{a_n^{(n)} a_{n-1}^{(n-1)} \ldots a_1^{(1)}\}^2 R(\theta_1, \theta_0)$, or, by using (2.8) and $R(\theta_1, \theta_0) = 1$,

(2) $$R(\theta_n, \theta_{n-1}) = \{\prod_{k=1}^{n-1} \frac{(2^k)!}{2^k k!}\}^2.$$

On the other hand, by (3.8), $\theta_n'(x_i) = -x_i \theta_{n-1}(x_i)$ and, if we take the product over all zeros x_i of $\theta_n(x)$, we obtain

$$R(\theta_n, \theta_n') = \prod_{i=1}^{n} \theta_n'(x_i) = (-1)^n x_1 \ldots x_n \prod_{i=1}^{n} \theta_{n-1}(x_i) = (-1)^n a_n^{(n)} R(\theta_n, \theta_{n-1}).$$

By (1), the first member equals $(-1)^{n(n-1)/2} D_n$ and it follows that $D_n = (-1)^{n(n+1)/2} a_n^{(n)} R_n(\theta_n, \theta_{n-1})$, so that, by (2),

(3) $$D_n = (-1)^{\frac{n(n+1)}{2}} \frac{(2n)!}{2^n n!} \{\prod_{k=1}^{n-1} \frac{(2k)!}{2^k k!}\}^2.$$

The denominator in (3) may be written as $(2.4.6\ldots 2n) \prod_{k=1}^{n-1} (2.4\ldots 2k)^2$ and a fairly easy counting atgument shows that each even integer 2m occurs exactly to the power 2n-2m+1. This is exactly the same power to which it occurs in the numerator, so that $|D_n|$ reduces to the product of the odd integers in the numerator. The odd factor 2m-1 ($1 \leq m \leq n$) occurs exactly once in (2n)!, and also once in each factorial (2k)! for $m \leq k \leq n-1$, so that 2(n-1-(m-1))+1= 2n-2m+1 is the exact power to which

2m-1 occurs in (3). It follows that

(4) $$\pm D_n = \prod_{m=1}^{n} (2m-1)^{2n-2m+1} = 1^{2n-1}3^{2n-3}\ldots(2n-3)^3(2n-1)^1.$$

(b) Set $n-3 = \frac{9}{8}x$ and recall from Chapter 11 that, for $x \geq 48$, there is a prime in the interval $x < p \leq \frac{9}{8}x$. It follows that there is a prime p that satisfies the inequalities

$$\frac{2n-1}{3} < \frac{8n}{9} - \frac{8}{3} = x < p \leq \frac{9}{8}x = n-3 < n-2,$$

provided that $n \geq 3 + \frac{9}{8} \cdot 48 = 57$. It is an easy exercise to verify that for all $14 \leq n \leq 56$ there also exists such a prime (e.g. for $n = 20$, $\frac{2.20-1}{3} = 13 < 17 < 20-2$)

Also for $n = 10$ we find $\frac{2n-1}{3} = \frac{19}{3} < 7 < 8 = n-2$. By (4), it follows that any such prime divides D_n to the exact power $2n-p \geq n+2 > n$. It also follows from (2.8), that $p|a_n^{(n)}$, $p^2\nmid a_n^{(n)}$. Indeed, $p < n-2 < n+1 < 2p < 2n < 3p+1$; however, 2n is even, so that the last inequality is equivalent to $2n < 3p$.

(c) By the Lemma, $\theta_n(x) \equiv x^p\theta_{n-p}(x) \pmod p$ for $(n+1)/2 \leq p < n$, a fortiori for $(2n-1)/3 < p < n-2$. Next, by (4), $|D_{n-p}| = 1^{2n-2p-1}3^{2n-2p-3}\ldots(2n-2p-3)^3(2n-2p-1)^1$; however, $p > 2n-2p-1$, because, by assumption, $3p > 2n-1$, and $p\nmid D_{n-p}$.

(d) All primes that divide (2n)! only once, divide D_n to an odd power, so that, except for $n = 1, D_n$ is not a perfect square.

With this, all conditions of Theorem A are verified and it follows that for $n \geq 14$ and also for $n = 10$, the Galois group G_n of the irreducible BP $\theta_n(x)$ is the symmetric group S_n.

5. For $n = 13, 12, 11$ and $n \leq 9$, no primes exist in $(2n-1)/3 < p < n-2$. For $n = 1$ $G_1 = S_1$, both reducing to the identity. For $n = 2$ and $n = 3$, the polynomials are irreducible and the discriminants are not squares; hence $G_2 = S_2$ and $G_3 = S_3$.

For the remaining values of n, we try to find primes $p = p_n$ that satisfy at least the weaker inequalities $(2n-1)/3 < p < n$. We find exactly one such prime for all $4 \leq n \leq 13$, $(n \neq 10)$, except for $n = 5$ and $n = 11$. Specifically, $p_4 = 3$, $p_6 = p_7 = 5$, $p_8 = p_9 = 7$, $p_{12} = p_{13} = 11$. We may recall from Chapter 11, that all BP involved are in fact irreducible, hence, the G_n's are transitive and, as also in each case $p_n > n/2$, the respective G_n's are primitive. By Theorem E, the G_n's have degrees of transitivity $\ell_n \geq n-p_n+1$, so that $\ell_4 \geq 2$, $\ell_6 \geq 2$, $\ell_7 \geq 3$, $\ell_8 \geq 2$, $\ell_9 \geq 3$, $\ell_{12} \geq 2$, and $\ell_{13} \geq 3$.

By Theorem C (2) there are no primitive groups of degree 4, except for A_4 and S_4, so that $G_4 \supset A_4$ and (recall (d), whence $G_4 \neq A_4$), $G_4 = S_4$.

By Theorem C (3), the only triply transitive groups of degree seven are A_7 and S_7, whence, as before, $G_7 = S_7$. For n = 13 we invoke Theorem C (1), observe that $p_1 = 5$ and $p_2 = 2p_1+1 = 11$ are both primes and conclude that G_{13}, with $n = 2p_1+3 = 13$, contains A_{13}, so that $G_{13} = S_{13}$.

6. The only values of n for which Theorem 1' has not yet been proved are n = 5,6,8,9,11, and 12.

In these six cases we shall use Theorem B and observe that we have

$$\theta_5(x) \equiv (x^3+x^2+4x+5)(x-2)(x-1) \pmod{17}$$

$$\theta_6(x) \equiv (x^3+4x^2-3x+6)(x^2+x-1)(x+3) \pmod{13}$$

(5)

$$\theta_8(x) \equiv (x^5+9x^4-7x^3+8x^2-7x+4)(x^2+2x+2)(x+6) \pmod{19}$$

$$\theta_9(x) \equiv (x^5+x^4-5x^3-x^2+7x+13)(x^3+12x^2-13x+3)(x+3) \pmod{29}$$

$$\theta_{11}(x) \equiv (x^7+39x^6+35x^5+121x^4+148x^3+60x^2+25x+53)(x^4+27x^3+14x^2+74x+23) \pmod{149}$$

$$\theta_{12}(x) \equiv (x^7+2x^6+27x^5+28x^4+30x^3+38x^2+15x+3)(x^5+76x^4+65x^3+63x^2+29x+15) \pmod{89}$$

In each of these six factorizations, the polynomial factors on the right are mutually incongruent and irreducible modulo the respective prime modulus. The first three factorizations were done by hand, by the present author, in 1949; $\theta_9(x)$ has been factored modulo 29 by M.Newman and K. Kloss on the IBM 704 at the National Bureau of Standards, in October 1961, while $\theta_{11}(x)$ and $\theta_{12}(x)$ have been factored by J.D. Brillhart and R. Stauduhar, on the CDC 6400 of the University of Arizona, in February 1969, by using an algorithm of E. Berlekamp.

It is, of course, quite easy to verify these factorizations by hand, once they are found. The discovery of a prime that leads to a usable factorization and the determination of the factors is perhaps less trivial. The verification that the factors are indeed irreducible modulo the respective prime is an operation of intermediate difficulty. For whatever interest it may present, the factorization of $\theta_8(x)$ will be discussed in detail in Section 7 and is fairly typical for all similar factorizations.

By Theorem B we know that for n = 5,6,8,9,11, and 12 the corresponding G_n contain at least one permutation P_n, of the following structures, respectively:

$P_5 = (a,b,c)$ $\qquad$ $P_9 = (a,b,c,d,e)(f,g,h)$

$P_6 = (a,b,c)(d,e)$ $\qquad$ $P_{11} = (a,b,c,d,e,f,g)(h,i,j,k)$

$P_8 = (a,b,c,d,e)(f,g)$ $\qquad$ $P_{12} = (a,b,c,d,e,f,g)(h,i,j,k,\ell)$.

P_5 leaves two symbols invariant; P_6, P_8, and P_9 each leave one symbol invariant, while P_{11} and P_{12} move all symbols.

G_5 contains a cycle of order three, hence the order of G_5 is divisible by 3; as $3 < (\frac{2}{3}) \cdot 5$, $3|5!$, and $3^2 \nmid 5!$, Theorem C (4) with e = 1 shows that $G_5 \supset A_5$, whence $G_5 = S_5$.

With P_6, G_6 contains also $P_6^3 = (e,d)$, a transposition. As a primitive group that contains transpositions, $G_6 = S_6$.

With P_8, G_8 also contains $P_8^2 = (a,c,e,b,d)$, an element of order 5. The order of G_8 is, therefore, divisible by 5 and, by Theorem C (4) with e = 1 and the remark that $5 < (\frac{2}{3}) \cdot 8$, it follows that $G_8 \supset A_8$, i.e. $G_8 = S_8$.

Similarly, G_9 contains P_9^3 of order $5 (< (\frac{2}{3}) \cdot 9)$, $5|9!$, $5^2 \nmid 9!$ and we conclude as before that $G_9 = S_9$. For G_{11} we use P_{11}^4 of order $7 (< (\frac{2}{3}) \cdot 11)$ and for G_{12} we use P_{12}^5 also of order 7 to reach the same conclusion. With this the proof of Theorem 1', hence also that of Theorem 1 and of Theorem 1", is complete.

7. In order to prove that $G_n \supset A_n$, the first step is to decide on the type of a desired factorization. For G_6 we could show that it contains a transposition, but in general, this cannot be achieved. In that case we try to find a $p_1 < 2n/3$, such that $p_1|n!$, $p_1^2 \nmid n!$, in order to apply Theorem C (4). For n = 8, there is little choice; only $p_1 = 5$ can be used, so that we aim at a factorization from which the existence of an element of order 5 can be inferred. Next, we have to choose a prime p as a modulus in Theorem B. No primes less than 2n can be used, because for them $a_0^{(n)} \equiv a_1^{(n)} \equiv 0 \pmod p$ and $\theta_n(x) \equiv x^2 g(x) \pmod p$ already exhibits a repeated factor, which makes Theorem B inapplicable. For n = 8, the smallest prime to be tried is, therefore p = 17. We first look for linear factors modulo 17, but there are none, because $\theta_8(k) \not\equiv 0 \pmod{17}$ for all integers k. Next one may want to consider a decomposition $\theta_8(x) \equiv f_2(x) f_6(x) \pmod{17}$, with a quadratic factor $f_2(x)$ and a factor $f_6(x)$ of degree 6. This, however, would not be useful. Indeed, in the

absence of linear factors one cannot obtain a factor of degree 5, by further factoring of $f_6(x)$. Hence, we attempt to split $\theta_8(x)$ as a product $f_3(x)f_5(x) \pmod{17}$. If we write monic polynomials f_3 and f_5, of degrees 3 and 5 respectively, with undetermined coefficients, we are led to a system of congruences modulo 17. It turns out that this system has no solution. As a factorization $\theta_4(x) \equiv f_4(x)\tilde{f}_4(x) \pmod{17}$ would not be useful, we pass to the next higher prime $p = 19$. Here we find that $\theta_8(-6) \equiv 0 \pmod{19}$ and, by factoring out $x+6$, we obtain

$$\theta_8(x) \equiv (x+6)(x^7-8x^6-6x^5-7x^4-5x^3+6x^2-6x+8) \pmod{19}. \tag{6}$$

By direct substitution of integers k $(k = 0,1,\ldots,18)$ into the factor of 7-th degree we verify that there exist no other linear factors. Hence, we look for a congruence of the form

$$x^7-8x^6-6x^5-7x^4-5x^3+6x^2-6x+8 \equiv (x^2+\alpha'x+\beta')(x^5+\alpha x^4+\beta x^3+\gamma x^2+\delta x+\varepsilon) \pmod{19}. \tag{7}$$

This leads to the following system of simultaneous congruences, all modulo 19:

$$\begin{array}{llll} \alpha+\alpha' \equiv -8 & \gamma+\alpha'\beta+\beta'\alpha \equiv -7 & \varepsilon+\alpha'\delta+\beta'\gamma \equiv +6 & \beta'\varepsilon \equiv 8. \\ \beta+\alpha'\alpha+\beta' \equiv -6 & \delta+\alpha'\gamma+\beta'\beta \equiv -5 & \alpha'\varepsilon+\beta'\delta \equiv -6 & \end{array}$$

This system has a solution modulo 19, which, when substituted in (6), leads to the third congruence of (5).

Before we can use Theorem B, we must still verify that all factors are irreducible (mod 19). As there are no linear factors, the only further possible decomposition would be

$$x^5+9x^4-7x^3+8x^2-7x+4 \equiv (x^2+\alpha'x+\beta')(x^3+\alpha x^2+\beta x+\gamma) \pmod{19}.$$

We are again led to a system of congruences (mod 19) in the undetermined coefficients and it turns out that this system has no solution. It follows that the fifth degree polynomial is irreducible modulo 19 and Theorem B may be applied to the factorization of $\theta_8(x)$ in (5).

CHAPTER 13

ASYMPTOTIC PROPERTIES OF THE BP

1. Asymptotic properties of BP were considered already in [53]. It was shown there that, for constant $z \neq 0$ and $n \to \infty$,

$$(1) \qquad y_n(z) \sim \frac{(2n)!}{2^n n!} z^n e^{1/z}.$$

Moreover, for $n > 1$,

$$(2) \qquad \left|y_n(z) - \frac{(2n)!}{2^n n!} z^n e^{1/z}\right| \le \frac{(2n)!}{2^{n+2} n!(n-1)} |z|^{n-2} e^{1/|z|} = k_n(z) \left|\frac{(2n)!}{2^n n!} z^n e^{1/z}\right|,$$

where $k_n(z) = \frac{1}{4(n-1)} \left|z^{-2} \exp(\frac{1}{|z|} - \frac{1}{z})\right|$ and $k_n(z) \to 0$ for fixed $z \neq 0$ and $n \to \infty$.

If one uses Stirling's formula for the factorials, (1) is seen to be equivalent to

$$(3) \qquad y_n(z) \sim (\frac{2nz}{e})^n \sqrt{2}\ e^{1/z}.$$

Formula (3) has been generalized by Obreshkov [82] to the polynomials $y_n(x;a,b)$ with $b = -1$. If in Obreshkov's formula one makes the change of variable $-2x = z$ needed to reduce b to its standard value $b = 2$, and recalls that in his notation $m = a-2$, the formula of [82] reads

$$(4) \qquad y_n(z) \sim (\frac{2nz}{e})^n 2^{a-3/2} e^{1/z}.$$

Formula (4) has been improved further by K. Dočev [48], who proved that

$$(5) \qquad y_n(z) = (\frac{2nz}{e})^n 2^{a-3/2} e^{1/z} \{1 - \frac{1+6(a-2)(a-1+2z^{-1})+6z^{-2}}{24n} + O(\frac{1}{n^2})\}.$$

It is clear from Dočev's work that the terms of the large bracket are only the first few terms of an asymptotic series, which, in principle, can be computed to any desired number of terms.

Formulae (1) to (5) are meaningful only for $z \neq 0$. In [53] the remark is made that close to $z = 0$, $y_n(z)$ behaves essentially like $e^{n^2 z/2}$, and even an error term is computed. In fact, it is clear from (2.10) and (2.8) that if one wants to approximate $y_n(z)$ in the vicinity of $z = 0$, then the natural candidate is $e^{n(n+1)z/2}$. Indeed, $y_n(z) = 1+(n(n+1)/2)z+((n-1)n(n+1)(n+2)/2.4)z^2 + O(z^3)$ and $e^{n(n+1)z/2} = 1+(n(n+1)/2)z+(n^2(n+1)^2/2.4)z^2 + O(z^3)$, so that an elementary computation leads to

THEOREM 1. *For fixed* n *and* $|z| \to 0$,

$$y_n(z) - e^{n(n+1)z/2} = -\frac{1}{4} n(n+1)z^2 + O(z^3);$$

in particular, $y_n(z) \sim e^{n(n+1)z/2}$.

This result is better than that of [53], but it seems that neither of them has any particular significance. Therefore we shall not consider them in what follows and turn to discuss the approximation of the exponential $e^{1/z}$ by $\frac{2^n n!}{(2n)!} y_n(z) z^{-n}$ (for $z \neq 0$), or, equivalently, that of e^z by $\frac{2^n n!}{(2n)!} \theta_n(z)$ (for z finite), as expressed by formulae (1) to (5). This approximation is precisely the property of BP used in the proof of the irrationality of e^r for r rational (see the Introduction, Problem 1, and Chapter 14).

2. It is, of course, sufficient to prove (5), from which the other formulae readily follow This proof, which is the object of Section 5, uses rather deep results, such as Dočev's estimate of the zeros of $y_n(z,a)$ (see Theorem 10.6). For this reason we present in sections 2 to 4 an independent, elementary, although somewhat lengthy , proof of the following slightly stronger version of (1) and (3).

THEOREM 2. *For constant* $z \neq 0$,

$$y_n(z) = c_0 z^n e^{1/z} + R_n(z),$$

with $c_0 = \frac{(2n)!}{2^n n!}$, *and where, for* $|z| \geq 2/n$, *one has* $|R_n(z)| < c_0 |z|^n \lambda(n,z)$ *with*

$0 \leq \lambda(n,z) \leq \frac{|z|^{-2} e^{1/|z|}}{2(2n-1)} (1+\mu(n))$; *here* $0 \leq \mu(n) \leq 2$ *and, for* $n \to \infty$,

$\mu(n) = \frac{8+\varepsilon_n}{\sqrt{2\pi}} n^{-3/2}$, *with* $\lim_{n \to \infty} \varepsilon_n = 0$.

REMARKS. From Theorem 2 one obtains (1), by the observation that $\lim_{n\to\infty} |R_n(z)/y_n(z)| = 0$. Also the error term estimate in (2) is an immediate consequence of the bound on $|R_n(z)|$ in Theorem 2. Finally, by Stirling's formula,

$$c_0 = \frac{(2n)!}{2^n n!} \sim \frac{(2n)^{2n+1/2} e^{-2n}}{2^n n^{n+1/2} e^{-n}} = 2^{n+1/2} e^{-n},$$

so that (3) follows from (1).

3. Proof of Theorem 2. In Chapter 2, by use of (2.10), we obtained the result

$$y_n(z) = c_0 \sum_{m=0}^{n} b_m^{(n)} z^{n-m}, \text{ with } c_0 = \frac{(2n)!}{2^n n!} \text{ and } b_m^{(n)} = \frac{2^m}{m!} \frac{(2n-m)!}{(n-m)!} \frac{n!}{(2n)!} .$$

Hence, $y_n(z) = c_0 z^n\{\sum_{m=0}^{n} \frac{z^{-m}}{m!} - \sum_{m=0}^{n}(1-\frac{2^m(2n-m)!n!}{(n-m)!(2n)!})\frac{z^{-m}}{m!}\}$

$$= c_0 z^n\{e^{1/z} - [\sum_{m=n+1}^{\infty}\frac{z^{-m}}{m!} + \sum_{m=0}^{n}(1-f(m,n))\frac{z^{-m}}{m!}]\}$$

$$= c_0 z^n e^{1/z} - c_0 z^n(R_1(z)+R_2(z)), \text{ say,}$$

where $f(m,n) = \frac{2^m n!(2n-m)!}{(2n)!(n-m)!}$. Also,

$$|R_1(z)| \le \frac{|z|^{-n-1}}{(n+1)!}(1+\frac{|z|^{-1}}{n+2}+\frac{|z|^{-2}}{(n+2)(n+3)}+\ldots) \le \frac{|z|^{-n-1}}{(n+1)!}\,\frac{1}{1-\frac{|z|^{-1}}{n+2}} < \frac{2|z|^{-n-1}}{(n+1)!}$$

for $|z| \ge 2/n$.

In order to estimate $|R_2(z)| = |\sum_{m=0}^{n}(1-f(m,n))\frac{z^{-m}}{m!}|$, we use the following

LEMMA 1. *For* $0 \le m \le n$,

$$0 \le 1 - f(m,n) \le \frac{m(m-1)}{2(2n-1)}, \tag{6}$$

with equality for $m = 0$ *and* $m = 1$; *the inequality on the left is strict for* $3 \le m \le n$ *and on the right for* $4 \le m \le n$.

If we accept for a moment (6), it follows that, for $z \neq 0$,

$$|R_2(z)| < \sum_{m=2}^{n}\frac{m(m-1)}{2(2n-1)}\,\frac{|z|^{-2}}{m!} =$$

$$\frac{|z|^{-2}}{2(2n-1)}\sum_{m=0}^{n-2}\frac{|z|^{-m}}{m!} \le \frac{|z|^{-2}}{2(2n-1)}\sum_{m=0}^{\infty}\frac{|z|^{-m}}{m!} = \frac{|z|^{-2}}{2(2n-1)}e^{1/|z|}.$$

If we set $R_3(z) = R_1(z) + R_2(z)$, then

$$|R_3(z)| \le \frac{2|z|^{-n-1}}{(n+1)!} + \frac{|z|^{-2}e^{1/|z|}}{2(2n-1)} = \frac{|z|^{-2}e^{1/|z|}}{2(2n-1)}\{1+\frac{4|z|^{-n+1}(2n-1)}{(n+1)!}e^{-1/|z|}\}.$$

For positive $u = |z|^{-1}$, the function $e^u u^{1-n}$ increases to infinity for $u \to \infty$ and also for $u \to 0$ ($n > 1$) and it attains its minimum for $u = n-1$, when it equals $(\frac{e}{n-1})^{n-1}$.

Consequently, $|R_3(z)| \le \frac{|z|^{-2}e^{1/|z|}}{2(2n-1)}(1+\mu(n))$, with $\mu(n) = \frac{4(2n-1)}{(n+1)!}(\frac{n-1}{e})^{n-1}$.

One easily verifies that, for $n \ge 1$,

$\mu(n+1) = \mu(n)\frac{(2n+1)n}{(2n-1)(n+2)}(\frac{n}{n-1})^{n-1}\frac{1}{e} < \frac{2n+1}{2n+3-2/n}\mu(n)$, so that, for $n \ge 2$,

$\mu(n+1) < \mu(n)$. Also, by direct computation $\mu(1)$ $(= \lim_{t \to 1}\frac{4(2t-1)}{\Gamma(t+2)}(\frac{t-1}{e})^{t-1}) = 2$,

$\mu(2) = 2/e$, $\mu(3) = 10/3e^2, \ldots$; hence, $\mu(n) \le 2$. Finally, by use of Stirling's formula, $\mu(n) \sim \frac{8}{\sqrt{2\pi}} n^{-3/2}$ for $n \to \infty$. It follows that $y_n(z) = c_0 z^n e^{1/z} + R_n(z)$, with $R_n(z) = c_0 z^n R_3(z)$ and

$$(7) \qquad |R_n(z)| \le c_0 \, |z|^n \, \frac{|z|^{-2} e^{1/|z|}}{2(2n-1)} \, (1+\mu(n)).$$

If we now compare the second member of (2) with (7), it follows that (2) is a consequence of Theorem 2, provided that $\frac{1}{4(n-1)} \ge \frac{1+\mu(n)}{2(2n-1)}$, or $\mu(n) < \frac{1}{2(n-1)}$. This obviously holds for large n, because $\mu(n) = O(n^{-3/2})$; in fact, this holds for any n, provided that $\frac{8(2n-1)(n-1)}{(n+1)!} \left(\frac{n-1}{e}\right)^{n-1} < 1$, as follows by using the explicit formula for $\mu(n)$. By direct computation one may check that this is indeed the case for all $n \ge 1$. The proof of (1), (2) and of Theorem 2 are complete, except for the proof of Lemma 1.

4. In the proof of Lemma 1 we shall need

LEMMA 2. *The inequality*

$$(8) \qquad \left(2n-1-\frac{m(m-1)}{2}\right)(2n-2)(2n-3)\ldots(2n-m+1) \le (2n-2)(2n-4)\ldots(2n-2m+2)$$

holds for $3 \le m \le n$ *and is strict for* $4 \le m \le n$.

Proof of Lemma 2. For $m(m-1) \le 4n-2$ the statement is trivial, because then the left hand side is non-positive, while the right hand side is positive. For $m = 3$, (8) reads $(2n-4)(2n-2) \le (2n-2)(2n-4)$ and reduces to an equality.

We now assume that (8) holds for some m with $3 \le m < n$ and show that it also holds for m+1. If we replace m by m+1, the first factor in (8) becomes $2n-1-\frac{(m+1)m}{2} = 2n-1-\frac{m(m-1)}{2}-m$, and we have to verify that

$$\left(2n-1-\frac{m(m-1)}{2}-m\right)(2n-2)(2n-3)\ldots(2n-m+1)\cdot(2n-m) \le (2n-2)(2n-4)\ldots(2n-2m+2)(2n-2m).$$

This is equivalent to

$$\left(2n-1-\frac{m(m-1)}{2}\right)(2n-2)(2n-3)\ldots(2n-m+1)\cdot(2n-m)-m(2n-2)(2n-3)\ldots(2n-m+1)(2n-m)$$
$$\le (2n-2)(2n-4)\ldots(2n-2m+2)(2n-2m).$$

By the induction assumption (8), it is sufficient to prove that

$$(2n-2)(2n-4)\ldots(2n-2m+2)\cdot(2n-m)-m(2n-2)(2n-3)\ldots(2n-m)$$
$$\le (2n-2)(2n-4)\ldots(2n-2m+2)\cdot(2n-2m),$$

or

$$(2n-2)(2n-4)\dots(2n-2m+2)\cdot m \le (2n-2)(2n-3)\dots(2n-m)\cdot m.$$

This clearly holds, because the number of factors is the same on both sides of the inequality and corresponding factors are either equal (namely the first and last), or are larger on the right, than on the left.

<u>Proof of Lemma 1.</u> For $m = 0$ and $m = 1$, $f(m,n) = 1$ and $\frac{m(m-1)}{2(2n-1)} = 0$, so that both equalities hold. For $m = 3$, (6) becomes $0 < 1-f(3,n) = \frac{3}{2n-1} = \frac{m(m-1)}{2(2n-1)}$.

To complete the proof by induction, we assume that $3 \le m-1 < n$ and that (6) holds for $m-1$. We show now that (6) also holds if $m-1$ is replaced by m. The inequality on the left is immediate, because $f(m,n) = \frac{2(n-m+1)}{2n-m+1} f(m-1,n) < f(m-1,n)$. The inequality on the right is equivalent to

$$1 - \frac{2^{m-1}(n-1)(n-2)\dots(n-m+1)}{(2n-1)(2n-2)\dots(2n-m+1)} = 1 - \frac{(2n-2)(2n-4)\dots(2n-2m+2)}{(2n-1)(2n-2)\dots(2n-m+1)} \le \frac{m(m-1)}{2(2n-1)}.$$

By clearing of denominators we obtain

$$(2n-1)(2n-2)\dots(2n-m+1)-(2n-2)(2n-4)\dots(2n-2m+2) \le \frac{m(m-1)}{2}(2n-2)(2n-3)\dots(2n-m+1),$$

or $(2n-1-\frac{m(m-1)}{2})(2n-2)(2n-3)\dots(2n-m+1) \le (2n-2)(2n-4)\dots(2n-2m+2)$, which holds on account of Lemma 2. This completes the proof of Lemma 1, and with it that of (1), (2), (3) and of Theorem 2.

5. In this section we shall give the proof of

<u>THEOREM 3.</u> (Dočev [48]). *For constant* $z \neq 0$ *and* $n \to \infty$,

$$y_n(z;a) = (\frac{2nz}{e})^n 2^{a-3/2} e^{1/z}\{1-\frac{1}{24n}-\frac{z^{-2}}{4n}-\frac{(a-2)(a-1+2z^{-1})}{4n}+O(\frac{1}{n^2})\}. \tag{9}$$

In fact, in order to illustrate the remark made after (5) we shall actually improve (9) to (14) with the better error term $O(n^{-3})$.

<u>Proof of Theorem 3.</u> From (2.27) and with the notations of Chapter 2,

$$y_n(z;a) = d_n^{(n)} \sum_{k=0}^{n} \{d_k^{(n)}/d_n^{(n)}\} z^k = d_n^{(n)} \sum_{k=0}^{n} c_{n-k}^{(n)} z^k,$$

where

$$d_n^{(n)} (= d_n^{(n)}(a)) = 2^{-n}(2n+a-2)^{(n)} = 2^{-n}\,\Gamma(2n+a-1)/\Gamma(n+a-1), \tag{10}$$

$$c_{n-k}^{(n)} (= d_k^{(n)}/d_n^{(n)}) = 2^{n-k}\binom{n}{k}\Gamma(n+k+a-1)/\Gamma(2n+a-1).$$

If $\alpha_j (= \alpha_j^{(n)})$ $(j = 1,2,\dots,n)$ are the zeros of $y_n(z;a)$, then

$$y_n(z;a) = d_n^{(n)} z^n P(z), \tag{11}$$

where

$$(12) \qquad P(z) = \prod_{j=1}^{n} (1-\alpha_j/z).$$

By (10) and Stirling's formula

$$d_n^{(n)} = 2^{-n} \frac{(2n+a-1)^{2n+a-3/2}}{(n+a-1)^{n+a-3/2}} \frac{e^{-2n-a+1}}{e^{-n-a+1}} \frac{1+(12(2n+a-1))^{-1}+(288(2n+a-1)^2)^{-1}+0(n^{-3})}{1+(12(n+a-1))^{-1}+(288(n+a-1)^2)^{-1}+0(n^{-3})}.$$

By routine (but lengthy) computations we obtain the following development, with error term $0(n^{-3})$:

$$d_n^{(n)} = 2^{n+a-3/2} n^n e^{-n}\{1 - \frac{6(a-1)(a-2)+1}{24n} + \frac{h(a)}{1152n^2} + 0(n^{-3})\}$$

where $h(a) = 12(a-1)(a-2)(3a+7)+1$. Next, by (12),

$$\log P(z) = \log \prod_{j=1}^{n} (1-\frac{\alpha_j}{z}) = -\sum_{j=1}^{n} \sum_{r=1}^{\infty} \frac{\alpha^r}{rz^r} = -\sum_{r=1}^{\infty} \frac{\sigma_r}{rz^r},$$

where $\sigma_r = \sum_{j=1}^{n} \alpha_j^r$. The values of the first few σ_r's have been computed in Chapter 10. Specifically, $\sigma_1 = \frac{-2n}{2n+a-2}$, $\sigma_2 = \frac{4n(n+a-2)}{(2n+a-2)^2(2n+a-3)}$,

$\sigma_3 = \frac{-8(a-2)(n+a-2)n}{(2n+a-2)^3(2n+a-3)(2n+a-4)}$. Also, by Theorem 6 of Chapter 10, $|\alpha_j| \leq \frac{2}{n-1+\mathrm{Re}\, a}$,

so that $|\alpha_j^r| = 0(n^{-r})$ and $\sigma_r = 0(n^{1-r})$. Consequently,

$$\left|\sum_{r=4}^{\infty} \frac{\sigma_r}{rz^r}\right| \leq c \sum_{r=4}^{\infty} \frac{n^{1-r}}{r|z|^r} \leq \frac{cn}{4} \sum_{r=4}^{\infty} \frac{1}{(n|z|)^r} = \frac{c}{4} \frac{1}{n^3|z|^4(1-(n|z|)^{-1})},$$

with a constant c, that depends only on a.

For fixed z and n sufficiently large,

$$1-|nz|^{-1} > 1/2, \text{ so that } \left|\sum_{r=4}^{\infty} \frac{\sigma_r}{rz^r}\right| \leq \frac{c}{2} \frac{1}{n^3|z|^4} = 0(n^{-3}).$$

In fact, this estimate holds uniformly in z, provided that $|z| \geq \eta$, for some fixed $\eta > 0$.

We conclude that

$$\log P(z) =$$

$$\frac{2n}{2n+a-2} \frac{1}{z} - \frac{2n(n+a-2)}{(2n+a-2)^2(2n+a-3)} \frac{1}{z^2} + \frac{8(z-2)(n+a-2)n}{3(2n+a-2)^3(2n+a-3)(2n+a-4)} \frac{1}{z^3} + 0(n^{-3})$$

$$= z^{-1} - (\frac{(a-2)z^{-1}}{2} + \frac{z^{-2}}{4})n^{-1} + (\frac{(a-2)^2 z^{-1}}{4} - \frac{z^{-2}}{8})n^{-2} + 0(n^{-3})$$

By exponentiation we obtain

(13) $P(z) =$

$$e^{1/z}\{1-(\frac{(a-2)z^{-1}}{2} + \frac{z^{-2}}{4})n^{-1} + (\frac{(a-2)^2 z^{-1}}{4} + \frac{[(a-2)^2-1]z^{-2}}{8} + \frac{(a-2)z^{-3}}{8} + \frac{z^{-4}}{32})n^{-2}+0(n^{-3})\}.$$

Finally, we substitute (13) in (11) and obtain

(14) $$y_n(z;a) = (\frac{2nz}{e})^n 2^{a-3/2} e^{1/z} g(z,a),$$

where

$$g(z,a) = 1-[(a-2)(a-1+2z^{-1})+(\frac{1}{6} + z^{-2})](4n)^{-1}$$
$$+ [(a-2)((a-1)(3a+7)+(12(a-2)(a+1)+2)z^{-1}+6(3a-5)z^{-2}+12z^{-3})$$
$$+ (\frac{1}{12} + 11z^{-2} + 3z^{-4})] (96n^2)^{-1} + 0(n^{-3}).$$

If we combine in (14) the terms of order n^{-2} with the term $0(n^{-3})$, then we obtain precisely Dočev's formula (9). In the particular case a = 2, (14) reduces to

$$y_n(z) = (\frac{2nz}{e})^n\sqrt{2}\, e^{1/z}\{1-(\frac{1}{6} +z^{-2})\frac{1}{4n} +(\frac{1}{12} + 11z^{-2} + 3z^{-4}) \frac{1}{96n^2} + 0(\frac{1}{n^3})\},$$

from which (1) and (3) immediately follow.

PART IV

CHAPTER 14

APPLICATIONS

1. In the present chapter we shall discuss some of the problems, in which BP have found applications and which, in fact, led to their study. Three of these have already been mentioned in the Introduction and sections 2-5 of the present chapter are mainly an expanded version of the earlier presentation. Somewhat unfortunately for this exposition, those problems present their own difficulties; while these can be overcome by classical, well established methods, even the short sketches of those procedures given here, may well divert the attention of the reader from the role played by the BP themselves in the solutions of the respective problems.

2. The irrationality of the exponential function e^c for rational c. While the irrationality of e is an almost immediate consequence of its usual series expansion, the irrationality of e^c for rational c is less obvious. We shall prove it here, by using results of Chapter 8.

We recall that the polynomials $P_n(x)$ collect the terms of $y_n(x)$ with powers of x of the same parity as n, while $Q_n(x)$ is the sum of the other terms of $y_n(x)$. Let x be a fixed, positive, rational number and set $x/2 = s/r$, with coprime integers s and r, so that $e^{2/x} = e^{r/s}$. It is clear that $r^n P_n(2s/r) = A_n$ and $r^{n-1}Q_n(2s/r) = B_n$ are integers. Also, $Q_n(x) = a_{n-1}^{(n)} x^{n-1} + a_{n-3}^{(n)} x^{n-3} + \ldots > a_{n-1}^{(n)} x^{n-1}$, so that, in particular, $Q_n(2s/r) > a_{n-1}^{(n)} 2^{n-1}(s/r)^{n-1} = \frac{(2n)!}{2(n!)} (\frac{s}{r})^{n-1}$ and $B_n = r^{n-1}Q_n(2s/r) > \frac{(2n)!}{2(n!)} s^{n-1}$. If $e^{r/s}$ would be rational, then $\frac{e^{r/s}+1}{e^{r/s}-1} = \frac{a}{b}$, with a,b coprime positive integers would follow. By Corollary 8.4

$$0 < \left|\frac{P_n(2s/r)}{Q_n(2s/r)} - \frac{a}{b}\right| = \left|\frac{A_n}{rB_n} - \frac{a}{b}\right| < \frac{1}{Q_n(2s/r)Q_{n+1}(2s/r)} = \frac{r^{2n-1}}{B_n B_{n+1}},$$

or

$$0 < |bA_n - arB_n| < \frac{r^{2n}b}{B_{n+1}} < \frac{r^{2n}(2b)}{s^n} \frac{(n+1)!}{(2n+2)!} \leq 2br^{2n} \frac{(n+1)!}{(2n+2)!}.$$

By Stirling's formula, the right hand term is less than $\frac{\sqrt{2}\, b}{er^2} (\frac{r^2 e}{4n})^{n+1} (1+\varepsilon_n) (\lim_{n\to\infty} \varepsilon_n = 0)$; hence, for sufficiently large n, the positive integer $|bA_n - arB_n|$ is less than one, which is impossible and shows that $e^{r/s}$ cannot be rational. In particular e itself and all its integral positive and negative powers are irrational.

3. The irrationality of π. We shall use again results of Chapter 8 and follow, in the main, the presentation of Siegel [53].

First we obtain a new representation for $R_n(x)$, defined by (8.6). We differentiate (8.6) n+1 times, observe that $\frac{d^{n+1}}{dx^{n+1}}\theta_n(x/2) = 0$, use Leibniz's formula for the derivative of a product and obtain

(1) $\frac{d^{n+1}}{dx^{n+1}} R_n(x) = -\frac{d^{n+1}}{dx^{n+1}}\{\theta_n(-x/2)e^x\} = \sum_{r=0}^{n+1}(-1)^{r+1}\binom{n+1}{r}2^{-r}\theta_n^{(r)}(-x/2)e^x = e^x S(x);$

here $S(x) = \sum_{j=0}^{n}\alpha_j x^j$ is a polynomial in x of degree at most n. We now observe that, on the one hand, by (8.6), $\frac{d^{n+1}}{dx^{n+1}}R_n(x)$ contains only powers of x with exponent at least n; on the other hand,

$e^x S(x) = \alpha_0+(\alpha_1+\alpha_0)x+\ldots+(\alpha_j+\alpha_{j-1}+\ldots+\alpha_0/j!)x^j+\ldots+(\alpha_n+\alpha_{n-1}+\ldots+\alpha_0/n!)x^n+\ldots$.

It follows that $\alpha_0 = \alpha_1 = \ldots = \alpha_{n-1} = 0$. As for α_n, only the term of (1) with $r = 0$ contributes to it and it is immediate from (1) that $\alpha_n = (-1)^{n+1}2^{-n}$, so that $S(x) = (-1)^{n+1}2^{-n}x^n$ and $\frac{d^{n+1}}{dx^{n+1}}R_n(x) = (-1)^{n+1}2^{-n}x^n e^x$. By an (n+1)-fold integration from 0 to x it now follows that $R_n(x) = \frac{(-1)^{n+1}}{2^n n!}\int_0^x (x-t)^n t^n e^t dt$. Finally, if we replace t by tx, we obtain the desired new representation

(2) $$R_n(x) = (-1)^{n+1}\frac{x^{2n+1}}{2^n n!}\int_0^1 (1-t)^n t^n e^{tx}dt.$$

For $x = i\pi$, in particular,

$$|R_n(i\pi)| \le \frac{\pi^{2n+1}}{2^n n!}\int_0^1 (1-t)^n t^n dt = \frac{\pi^{2n+1}}{2^n n!}B(n+1, n+1),$$

where $B(u,v) = \frac{\Gamma(u)\Gamma(v)}{\Gamma(u+v)}$ is Euler's Beta function. Consequently,

$$|R_n(i\pi)| \le \frac{\pi^{2n+1}}{2^n}\frac{n!}{(2n+1)!}$$, or, by Stirling's formula,

(3) $$|R_n(i\pi)| \le \frac{\pi^{2n+1}}{2^n}\left(\frac{e}{4n}\right)^{n+1}\frac{1+\varepsilon_n}{e\sqrt{2}} = \left(\frac{\pi^2 e}{8n}\right)^{n+1}\frac{\sqrt{8}\,(1+\varepsilon_n)}{\pi e} \quad (\lim_{n\to\infty}\varepsilon_n = 0).$$

Let us assume that $\pi^2/4 = a/b$ is rational, with a,b coprime positive, rational integers. By setting $x = \pi i$ in (8.6), we obtain, with $\nu = [n/2]$,

$$R_n(i\pi) = \theta_n(\pi i/2)-\theta_n(-\pi i/2)e^{\pi i} = \theta_n(\pi i/2)+\theta_n(-\pi i/2) = 2\sum_{k=0}^{\nu} a_{n-2k}^{(n)}(\pi i/2)^{2k} =$$

$$2\sum_{k=0}^{\nu}(-1)^k a_{n-2k}^{(n)}(\pi/2)^{2k} = 2\sum_{k=0}^{\nu}(-1)^k a_{n-2k}^{(n)}(a/b)^k,$$

with $a_{n-2k}^{(n)}$ given by (2.8). By Theorem 3.1 it follows that $|b^{\nu}R_n(i\pi)| =$
$|2 \sum_{k=0}^{\nu} (-1)^k a_{n-k}^{(n)} a^k b^{\nu-k}| = N$ is a non-negative integer. In fact, $N > 0$, because (2) shows that for $x = \pi i$ the integrand has the positive imaginary part $(1-t)^n t^n \sin \pi t$ $(0 < t < 1)$. On the other hand, by (3),

$$0 < N = |b^{\nu}R_n(i\pi)| \leq b^{n/2}(\frac{\pi^2 e}{8n})^{n+1} \frac{\sqrt{8}\,(1+\varepsilon_n)}{\pi e} \leq (\frac{\pi^2 e b^{1/2}}{8n})^{n+1} \sqrt{\frac{8}{b}}\,\frac{1+\varepsilon_n}{\pi e} < 1$$

for sufficiently large n and this is not possible for any integer N. This shows that the assumption of the rationality of π^2 is not tenable, π^2 is irrational and this holds, of course, for π itself.

4. Solution of the wave equation in spherical coordinates. This problem is, of course, classical and the reader may want to consult, e.g. [55] for a traditional solution. The equation to be solved is

$$\Delta u = \frac{1}{c^2} \frac{\partial^2 u}{\partial t^2}, \tag{4}$$

where Δ is the Laplacian and c is the speed of propagation of the waves. In spherical coordinates

$$\Delta = \frac{\partial^2}{\partial r^2} + \frac{2}{r} \frac{\partial}{\partial r} + \frac{1}{r^2} (\frac{\partial^2}{\partial \theta^2} + \cot \theta \frac{\partial}{\partial \theta}) + \frac{1}{r^2 \sin^2 \theta} \frac{\partial^2}{\partial \phi^2} .$$

It is sufficient to consider only monochromatic waves, i.e., waves of a single frequency; indeed, the general case is reducible to the monochromatic one by ordinary Fourier analysis, that means by superposition of a countable number of monochromatic waves.

In order to solve (4), we use the method of separation of variables, i.e., we set

$$u = R(r)\Theta(\theta)\Phi(\phi)T(t), \tag{5}$$

where each factor depends only on a single independent variable. Specifically, r is the distance of a point P from the origin O, θ is the angle of OP with a fixed "polar" axis $(0 \leq \theta \leq \pi)$, ϕ is the angle that a plane through P and the polar axis makes with a fixed plane through the polar axis, and t is the time variable.

If we substitute this in (4) and divide by u, we obtain

$$\frac{R''}{R} + \frac{2}{r}\frac{R'}{R} + \frac{1}{r^2}\frac{\Theta'}{\Theta} \cot \theta + \frac{1}{r^2} \frac{\Theta''}{\Theta} + \frac{1}{r^2 \sin^2 \theta} \frac{\Phi''}{\Phi} = \frac{1}{c^2} \frac{T''}{T} . \tag{6}$$

Here the right hand member depends only on the time t, while the left hand member stays unchanged when t varies; hence, both sides of (6) are constant. In order to obtain a periodic solution, this constant must be negative and we denote it by $-k^2$.

Hence, $T'' + k^2c^2T = 0$ and, with proper choice of the origin for time, and after the suppression of a multiplicative constant (that would anyhow be absorbed into a product of similar constants - see (5)) we obtain $T = \cos kct$. We proceed in the same way with the separation of the other variables. First, we obtain

$$(7) \qquad r^2\sin^2\theta\{\frac{R''}{R} + \frac{2}{r}\frac{R'}{R} + \frac{\Theta''}{\Theta} + \frac{1}{r^2}\cot\theta\cdot\frac{\Theta'}{\Theta} + k^2\} = -\frac{\Phi''}{\Phi} = \mu.$$

In order to obtain a solution $\Phi(\phi)$ that is single valued and periodic with period 2π (as it has to be), it is necessary to take $\mu = q^2$, with q an integer, so that $\Phi(\phi) = a_q \cos q\phi + b_q \sin q\phi$. Next, (7) with $\mu = q^2$ leads to

$$(8) \qquad r^2\frac{R''}{R} + 2r\frac{R'}{R} + k^2r^2 = -\{\frac{\Theta''}{\Theta} + \cot\theta\frac{\Theta'}{\Theta} + \frac{q^2}{\sin^2\theta}\} = \lambda.$$

To solve the last (ordinary) differential equation of (8), set $\cos\theta = x$, $\Theta(\theta) = F(\cos\theta) = F(x) = (1-x^2)^{q/2}V(x)$. Then $V(x)$ satisfies the differential equation

$$(9) \qquad (1-x^2)V'' - 2(q+1)xV' + (\lambda-q^2-q)V = 0.$$

Equation (9) has, in general, only solutions by (often divergent) series; it admits, however, polynomial solutions if and only if $\lambda = n(n+1)$, for some integer n. In that case, its solution is $V(x) = \frac{d^q}{dx^q}L_n(x)$, where $L_n(x)$ is the Legendre polynomial of degree n. The proof is immediate, by q-fold differentiation of the differential equation for Legendre polynomials:

$(1-x^2)L_n''-2xL_n'+n(n+1)L_n = 0$. It follows that $\Theta(\theta) = \sin^q\theta\{\frac{d^q}{dx^q}L_n(x)\}_{x = \cos\theta}$, a function usually denoted by $P_n^q(\cos\theta)$, the associated Legendre function. It is clear that for $q > n$, $P_n^q(\cos\theta) = 0$ identically. We now solve the remaining ordinary differential equation. By (8) with $\lambda = n(n+1)$, this is

$$(10) \qquad r^2R'' + 2rR' + (k^2r^2-n(n+1))R = 0.$$

Equation (10) may be transformed into Bessel's equation. We prefer, however, to proceed differently. In order to obtain polynomial solutions we set $x = (ikr)^{-1}$ and $R(r) = (kr)^{-1}e^{-ikr}y(1/ikr) = ixe^{-x}y(x)$, so that (10) becomes

$$(11) \qquad x^2y'' + 2(x+1)y' - n(n+1)y = 0.$$

In (11) we recognize equation (2.11) of the BP $y_n(x)$. The result so far is that (4) admits solutions of the type

(12) $u(r,\phi,\theta,t) = (ikr)^{-1}e^{-ikr}y_n(1/ikr)\ (a_q \cos q\phi+b_q \sin q\phi)P_n^q(\cos\theta)\cos(kct)$,

a fact that can be verified by direct substitution of (12) in (4), or (5).

With (12) also

$$U(r,\theta,\phi,t) = \sum_{n=0}^{\infty}(ikr)^{-1}e^{-ikr}y_n(1/ikr)\sum_{q=0}^{n}(a_{n,q}\cos q\phi+b_{n,q}\sin q\phi)P_n^q(\cos\theta)\cos(ckt)$$

are solutions, for any choice of the constants $a_{n,q}$ and $b_{n,q}$. One can use this fact, in order to choose the constants so as to satisfy initial and boundary conditions.

Let us assume that at $t = 0$, $U(r,\theta,\phi,0) = F(r,\theta,\phi)$ is given. Then, if we set $(irk)^{-1} = x$ and $x^{-1}F((ikx)^{-1},\theta,\phi) = F_1(x,\theta,\phi)$, we obtain

(13) $$\sum_{n=0}^{\infty} e^{-1/x}y_n(x)\sum_{q=0}^{n}(a_{n,q}\cos q\phi+b_{n,q}\sin q\phi)P_n^q(\cos\theta) = F(x,\theta,\phi).$$

We now multiply both sides of (13) by $y_m(x)e^{-1/x}$ and integrate around the unit circle. By taking into account Corollaries 4.5 and 4.7 we obtain

$$(-1)^{m+1}\frac{2}{2m+1}\sum_{q=0}^{m}(a_{m,q}\cos q\phi+b_{m,q}\sin q\phi)P_m^q(\cos\theta) = \Psi_m(\theta,\phi),$$ where

$$\Psi_m(\theta,\phi) = \frac{1}{2\pi i}\int_{|x|=1}F_1(x,\theta,\phi)y_m(x)e^{-1/x}dx.$$ We rewrite this as

$$\sum_{q=0}^{m}(a_{m,q}\cos q\phi+b_{m,q}\sin q\phi)P_m^q(\cos\theta) = K_m(\theta,\phi),$$ where $K_m(\theta,\phi) = (-1)^{m+1}(m+1/2)\Psi_m(\theta,\phi)$. By using the orthogonality of the trigonometric functions and of the associate Legendre functions, one obtains (see, e.g. [55]) explicit values for $a_{m,q}$ and $b_{m,q}$:

$$4\pi a_{m,0} = (2m+1)\int_0^{2\pi}\int_0^{\pi}K_m(\theta,\phi)L_m(\cos\theta)\sin\theta\, d\theta d\phi;$$ and for $q \neq 0$,

$$2\pi a_{m,q} = (2m+1)\frac{(m-q)!}{(m+q)!}\int_0^{2\pi}\int_0^{\pi}K_m(\theta,\phi)P_m^q(\cos\theta)\cos q\phi\sin\theta\, d\theta d\phi,$$

$$2\pi b_{m,q} = (2m+1)\frac{(m-q)!}{(m+q)!}\int_0^{2\pi}\int_0^{\pi}K_m(\theta,\phi)P_m^q(\cos\theta)\sin q\phi\sin\theta\, d\theta d\phi.$$

Here $L_m(\cos\theta) = P_m^0(\cos\theta)$ is the Legender polynomial of degree m. Replacing again m by n, the function

(14) $$U(r,\theta,\phi,t) = \sum_{n=0}^{\infty}(ikr)^{-1}e^{-ikr}y_n(1/ikr)\sum_{q=0}^{n}(a_{n,q}\cos q\phi+b_{n,q}\sin q\phi)P_n^q(\cos\theta)\cos(kct)$$

is the required solution for the wave component corresponding to the given k. If there are several frequencies present, then the superposition of solutions of the type (14) will give the complete solution.

5. The infinite divisibility of certain probability distributions.

Let f(x) be a probability density function. Then it is customary in probability theory to call its Fourier transform $\phi(t) = \int_{-\infty}^{\infty} f(x)e^{itx}dx$, the "characteristic function" of f(x). Given a probability density f(x) and its characteristic function $\phi(t)$, let us set $\phi_m(t) = \{\phi(t)\}^{1/m}$. The question arises: is it true that for every integer m, $\phi_m(t)$ is the characteristic function of some probability distribution? In other words, given any integer m, is it true that there exists a non-negative, Lebesgue integrable function g(x) (= $g_m(x)$), such that $\phi_m(t) = \int_{-\infty}^{\infty} g(x)e^{itx}dx$? If such a function g(x) exists for every m, we say that the probability density function f(x) is infinitely divisible. The same terminology is used in the case of a (not necessarily differentiable) distribution function F(x). For a more detailed discussion of this topic see [17]. Much work has been done on the investigation of conditions that insure infinite divisibility (see, in particular, [44] and [19]). Recently, Thorin [58] and Bondesson [7] have determined large classes of distribution functions that are infinitely divisible. In fact, the results of [7] contain as particular cases those presented here; however, due to the generality of the problem considered in [7], the methods used there are rather deep. Here we shall restrict ourselves to some simple, specific cases, namely the Student t-distributions of an odd number of degrees of freedom. The corresponding density function is

$$f_{2k-1}(x) = c_k\left(1+\frac{x^2}{2k-1}\right)^{-k} \tag{15}$$

with the normalizing constant $c_k = \dfrac{\Gamma(k)}{\sqrt{(2k-1)\pi}\;\Gamma(k-1/2)}$.

REMARK. Let us observe parenthetically that if χ_m^2 is a chi-square random variable with m degrees of freedom, then the infinite divisibility of $f_m(x)$ implies also that of $(\chi_m^2)^{-1}$; we shall prove this infinite divisibility, as stated, for all odd $m = 2k-1$. A proof valid for all m (in fact, for all real m > 0) can be found in [56]. Still more general results are obtained in [36].

The present proof is based on the following Theorem A of Kelker [65]. We recall that a function $\phi(t)$ is said to be completely monotonic on an interval I, if on I the function and all its derivatives satisfy the inequality $(-1)^n\phi^{(n)}(t) \geq 0$ $(n = 0,1,2,\ldots)$.

THEOREM A. *The Student* t-*distribution of* 2m *degrees of freedom is infinitely divisible, if and only if the function* $\phi_m(t) = \frac{K_{m-1}(\sqrt{t})}{\sqrt{t}\,K_m(\sqrt{t})}$ *is completely monotonic on* $[0,\infty)$.

Here $K_m(u)$ is the Bessel (or Macdonald) function discussed in Chapter 2.

Kelker [65] and Ismail and Kelker [62] proved the complete monotonicity of $\phi_m(t)$ for some small half-odd integral values of m, from which the infinite divisibility of $f_{2k-1}(x)$ as given by (15) followed for the first few integral values of k. They also conjectured that $\phi_m(t)$ is completely monotonic for all real m, from which follows, in particular, the infinite divisibility of $f_{2k-1}(x)$ for all integers k. For further results on this problem, see besides [55] and [56], also [35], [36].

It is clear that if $\phi(f) = \int_0^\infty g(u)e^{-tu}du$, with $g(u) \geq 0$ for $0 < u < \infty$, then $\phi(t)$ is completely monotonic. That this sufficient condition is (essentially) also necessary is less obvious and is contained in the following theorem of S. Bernstein:

THEOREM B. (S. Bernstein; see [68]). *The function* $\phi(t)$ *is completely monotonic, if and only if it is the Laplace transform of a (not necessarily finite) measure, i.e. if and only if it is of the form* $\phi(t) = \int_0^\infty e^{-tu}dF(u)$.

In particular, if F(u) is differentiable, so that $F'(u) = g(u)$ exists, then one has $\phi(t) = \int_0^\infty g(u)e^{-tu}du$, with $g(u) \geq 0$ on $0 < u < \infty$. We shall denote the Laplace transform of g(u) by $L(g)(t)$ and the inverse transform of $\phi(t)$ by $L^{-1}(\phi)(u)$. We shall prove

THEOREM 1. *For every odd integer* $n \geq 1$, *set* $\phi(t) = \theta_{n-1}(t^{1/2})/\theta_n(t^{1/2})$ *and let* β_j $(= \beta_j^{(n)})$ $(j = 1,2,\dots,n)$ *be the zeros of the BP* $\theta_n(z)$; *then, if* $g(u) = L^{-1}(\phi)(u)$, *we have*

$$g(u) = (\pi u)^{-1/2} - 2\pi^{-1/2}\sum_{j=1}^{n} e^{\beta_j^2 u}\int_{-\beta_j\sqrt{u}}^{\infty} e^{-v^2}dv, \tag{16}$$

and g(u) *is completely monotonic; in particular,* $g(u) \geq 0$ *on* $0 < u < \infty$.

COROLLARY 1. *For every odd integer* n, $f_n(x)$ (as defined by (15) with n = 2k-1) *and* $(\chi_n^2)^{-1}$ *are infinitely divisible.*

Proof of Corollary 1. By (5.1), $K_{n+1/2}(z) = (\pi/2z)^{1/2}z^{-n-1/2}e^{-z}\theta_n(z)$; hence

$$\frac{K_{n-1/2}(z)}{zK_{n+1/2}(z)} = \frac{\theta_{n-1}(z)}{\theta_n(z)}$$

and

$$\phi_{n+1/2}(t) = \frac{K_{n-1/2}(t^{1/2})}{t^{1/2}K_n(t^{1/2})} = \frac{\theta_{n-1}(t^{1/2})}{\theta_n(t^{1/2})} .$$

Now the Corollary follows from Theorem 1 on account of Theorem A and the Remark preceding it. It remains to prove Theorem 1.

Proof of Theorem 1. By (10.14), $\dfrac{\theta_{n-1}(t^{1/2})}{\theta_n(t^{1/2})} = \sum_{j=1}^{n} \dfrac{1}{\beta_j(\beta_j - t^{1/2})}$; hence,

$$g(u) = L^{-1}(\phi)(u) = \sum_{j=1}^{n} \frac{1}{\beta_j} L^{-1} \frac{1}{\beta_j - t^{1/2}} = - \sum_{j=1}^{n} \frac{1}{\beta_j} L^{-1} \frac{1}{t^{1/2}+(-\beta_j)} .$$

By Theorem 10.5 and Corollary 10.2, Re $\beta_j < 0$; hence, $\mathrm{Re}(-\beta_j) > 0$ and the last inverse Laplace transform may be computed. In fact, it is known (see e.g., (29.3.37) in [1]) and we find

$g(u) = - \sum_{j=1}^{n} \frac{1}{\beta_j} \{\frac{1}{\sqrt{\pi u}} + \beta_j e^{\beta_j^2 u} \operatorname{erfc}(-\beta_j u^{1/2})\}$, where $\operatorname{erfc}(-\beta_j u^{1/2}) =$

$2\pi^{-1/2} \int_{-\beta_j\sqrt{u}}^{\infty} e^{-v^2} dv$. By using also (10.2') we obtain (16). It remains to show the complete monotonicity of g(u). In fact, we only need the positivity of g(u), in order to apply Theorem B, but it is easier to prove the stronger statement. Indeed, let

$$\Phi(x) = \pi^{-1}\{x^{-1/2} + \sum_{j=1}^{n} \beta_j x^{-1/2}(x+\beta_j^2)^{-1}\}.$$

We verify by direct computation, or on hand of tables of Laplace transforms (see, e.g., [1], (29.3.4) and (29.3.114)) that $L(\Phi)(u) = g(u)$. If we set $\Psi(x) = \pi x^{1/2}\Phi(x)$, then by Theorem B, g(u) is completely monotonic, provided that

$\Psi(x) = 1 + \sum_{j=1}^{n} \beta_j (x+\beta_j^2)^{-1} \geq 0$. Now, $\psi(x) = p(x)/q(x)$, with $q(x) = \prod_{j=1}^{n} (x+\beta_j^2)$, a polynomial with real coefficients.

As just recalled, we know from Chapter 10 that none of the β_j's is purely imaginary; hence, q(x) has no positive zeros and, consequently, does not change sign on $[0,\infty)$. As $q(0) = (\prod_{j=1}^{n} \beta_j)^2 = \{(2n)!/2^n n!\}^2 > 0$, one has $q(x) > 0$ for $0 \leq x < \infty$. Also

$$p(x) = q(x) + \sum_{j=1}^{n} \beta_j \prod_{k\neq j} (x+\beta_k^2) = x^n + \sum_{j=1}^{n} \gamma_j x^{n-j} \tag{17}$$

is a polynomial of exact degree n; it also has only real coefficients (namely symmetric functions of the β_j's) and so is real for real x. For $x \to \infty$, $\psi(x) = p(x)/q(x) \to 1$, so that, by $q(x) > 0$, also $p(x) > 0$ for large x. We now claim the validity of

LEMMA 1. $p(x) = x^n$.

Assuming the Lemma for a moment, $p(x) > 0$ for $0 < x < \infty$; hence, $\psi(x) > 0$ on $0 < x < \infty$, so that g(u) is indeed completely monotonic and the proof of Theorem 1 is complete. It only remains to justify the Lemma.

Proof of Lemma 1. From $q(0) \neq 0$ follows that $\psi(x)$ and $p(x)$ vanish (if at all) to the same order at $x = 0$. However, by (10.2') $\psi(0) = 1 + \sum_{j=1}^{n} \beta_j^{-1} = 0$ and $\psi^{(m)}(0) = \sum_{j=1}^{n} \beta_j^{-(2m+1)} = 0$ $(m = 1,2,\ldots,n-1)$, so that $\psi(x)$ has a zero of order at least n at $x = 0$. This shows that p(x), of degree n, also has a zero of order n at $x = 0$, so that in (17) all $\gamma_j = 0$ $(j = 1,2,\ldots,n)$ and $p(x) = x^n$, as claimed. With this the Lemma is proved and all assertions of this sections are justified.

6. Electrical networks. Let us consider a (perhaps very complex) electrical network. It can be decomposed into a finite collection of loops, each one of which may contain resistors, coils (self-inductances) and condensers (capacitors). If a closed loop contains no source of electrical potential, then the Ohm-Kirchhoff laws state that the total difference of potential around the loop vanishes. We recall that if the current $i = i(t)$ is measured in amperes then the difference of potential between the terminals of a resistor of R ohms equals $Ri(t)$ volts, while that between the terminals of a coil of self inductance of L henrys equals $L \frac{di(t)}{dt}$ volts, and that between the terminals of a condenser of capacity C equals $C^{-1} \int i(t)dt$ volts. If the loop is closed by a source of potential of v(t) volts, then the total drop of potential around the loop equals v(t).

To "solve" the network means to determine the currents in all branches and the potential (voltage) at all terminals (or "nodes"), with respect to some fixed potential, usually the "ground". In order to do that it is necessary to know also the initial conditions, i.e., the value of one such current and of such a difference of potential at, say, $t = 0$. We assume that in each loop, R, L, and C are "lumped", i.e., concentrated, rather than distributed and that they are constant in time. We consider, in particular, "passive" networks, that is networks without internal sources of energy. They will have, in general (see Fig. 1) two pairs of terminals ("ports"). To the first one we may connect an outside, known source, that applies there a given, variable potential $e_1(t)$, the input signal. The second port may be connected, e.g., to a large resistor, say ρ, and we are interested in the output potential $e_2(t)$ between its terminals, or, equivalently, in the output current $i_2(t) = e_2(t)/\rho$.

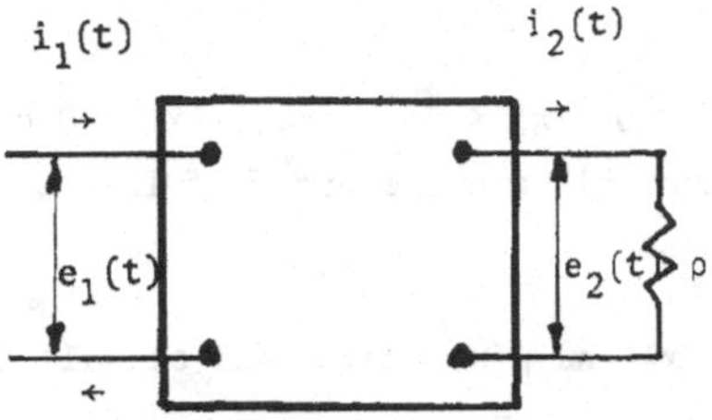

Fig. 1

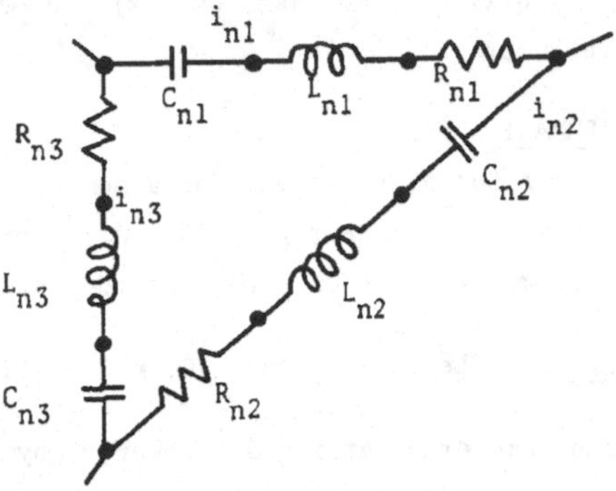

Fig. 2

A typical loop, say, the n-th loop may look like Fig. 2.

If we write the Ohm-Kirchhoff equations for the n-th loop, this has the form of an integro-differential equation

$$\text{(18)} \qquad \sum_j \{R_{nj} i_{nj}(t) + L_{nj} \frac{d}{dt} i_{nj}(t) + \frac{1}{C_{nj}} \int i_{nj}(t)dt\} = e_n(t)$$

($e_n(t) = 0$ if the loop is passive).

In order to solve this system, we replace each equation by its Laplace transform. This means that we multiply each term by e^{-st} ($s = \sigma + i\omega$, a complex variable) and integrate with respect to t (= time) from $t = 0$ to $t = \infty$. In this way each integro-differential equation becomes an algebraic equation in the new variable s. We set $I_{nj}(s) = \int_0^\infty i_{nj}(t)e^{-st}dt$ (whenever possible we shall suppress the subscripts) for the transform of a given current $i(t) = i_{nj}(t)$ in a given branch j of the n-th loop and $E(s) = \int_0^\infty e(t)e^{-st}dt$ for the transform of a difference of potential $e(t)$, between two specified points (or nodes, or terminals) of the network. A term $Ri(t)$ becomes $RI(s)$, $Li'(t)$ becomes $sLI(s)$, and $C^{-1} \int i(t)dt$ becomes $(Cs)^{-1}I(s)$. For the n-th loop, the transform of equation (18) becomes

$$\text{(19)} \qquad \sum_j \{R_{nj} + sL_{nj} + (sC_{nj})^{-1}\} I_{nj}(s) = E_n(s) \quad (E_n(s) = 0 \text{ if } e_n(t) = 0).$$

The problem of "synthesis" of a network that we address now consists in divising one to be put into the "black box" of Fig. 1, connected to the four terminals, so that, given a certain type of input, we should obtain a desired type of output. So, e.g., we may want that, if at the input we apply a large number of superposed potentials, each one oscillating with a different frequency, all but those within a given, narrow band should be suppressed, while those within that band should be collected at the output with a minimum of distortion. For instance, at the input we may collect through an antenna the e(t)'s due to many broadcasts, but at the output we want to eliminate all, except one, and this one with as little distortion as possible. Such a network, called a filter, would give us, ideally, an output like that of Fig. 3, with the "cut-off frequencies" ω_1 and ω_2. Often, with no particular loss of generality, we may set $\omega_1 = 0$ and then speak of a low-pass filter. Sometimes the purpose is to reproduce the input as faithfully (distortionless) as possible at the output. Then we speak of a time-delay network.

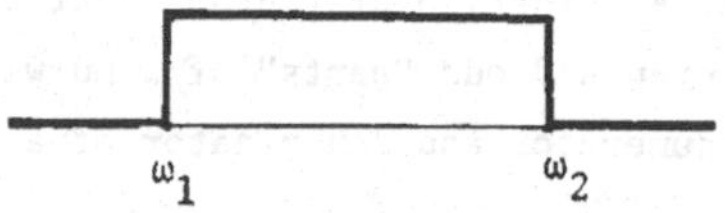

Fig. 3

In view of the linearity of the Ohm-Kirchhoff equations and of the Laplace transforms, also the relations between the transformed input and output currents and potentials, $I_1(s)$, $I_2(s)$, $E_1(s)$ and $E_2(s)$ are linear. In case one connects at the output a large resistor, in order only to collect there $e_2(t)$, the network may be considered a one-port (two-terminal) one. The relation between $I_1(s)$ and $E_1(s)$ being linear, there exists a function Z(s), called driving point impedance function, such that $E_1(s) = Z(s)I_1(s)$.

Similarly, $E_2(s)$ and $E_1(s)$ are related by $E_2(s) = T(s)E_1(s)$ (T(s) = transfer function). Both, Z(s) and T(s) are obtained as solutions of linear equations of the form (19), so that both are rational functions of s. In many cases one is interested in Z(s), or in T(s), for $s = i\omega$ ($\omega = 2\pi f$, f = frequency, $T = f^{-1}$ = period) purely imaginary; in that case one speaks of $Z(i\omega)$ simply as the (complex) impedance.

The following properties of Z(s) may be proved (see, e.g. [60]): If the network is passive and consists only of resistors, coils and condensers, then Z(s) is real for real s, $\mathrm{Re}\, Z(\sigma+i\omega) \geq 0$ for $\sigma \geq 0$, with $\mathrm{Re}\, Z(\sigma+i\omega) = 0$ possible only for $\sigma = 0$. Any function with these properties is said to be a p.r. (positive, real) function; it maps the closed right half-plane into itself, so that any purely imaginary boundary points of the image can have as inverse images only points also on the imaginary axis. If Z(s) is p.r., then so is $Z(s)^{-1}$ and also the composition of Z(s) with any other p.r. function. Neither zeros, nor poles of Z(s) may belong to the open right half-plane and if a pole is purely imaginary, then it must be simple, with positive residue. The difference between the degrees of numerator and denominator of Z(s) cannot exceed one. Many of these properties are shared by T(s), but not the last restriction.

If a polynomial has the property that it has no zeros in the right half-plane, it is called a Hurwitz polynomial. By what precedes, both, the numerator and the denominator of Z(s) (and also of T(s)) must be Hurwitz polynomials. The following Proposition holds:

PROPOSITION 1. *The ratio of the sum of the even powers to the sum of the odd powers of a Hurwitz polynomial is a p.r. function.*

The converse of this proposition is not quite true; indeed, if H(s) is a Hurwitz polynomial, then $K(s) = (s^2-a^2)H(s)$ is not Hurwitz; nevertheless, the ratio of the sum of its even powers to that of its odd powers is the same as for H(s), and hence is a p.r. function. We also remark that, after the elimination of factors that are common to the sums of even and of odd powers, the even and odd "parts" of a Hurwitz polynomial are themselves Hurwitz polynomials, as numerator and denominator of a p.r. function.

If we now consider the ideal output like the one in Fig. 3, it is rather clear that continuous functions will not lead to it. We may try to obtain an acceptable approximation to it by continuous functions, by settling for a graph of one of the two shapes of Fig. 4(a), or (b).

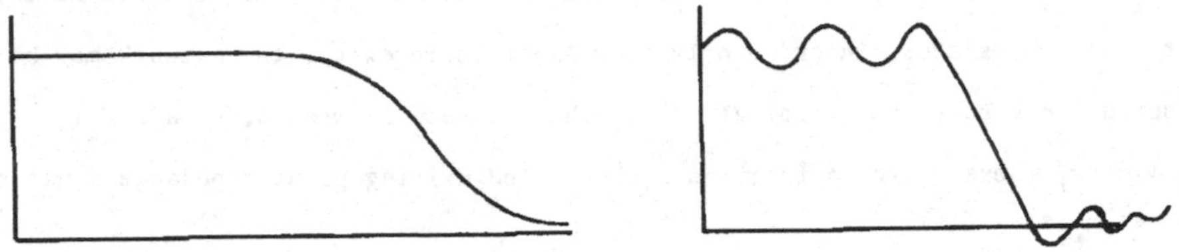

(a) (b)

Fig. 4

We face, however, still another difficulty. As seen, passive networks, consisting only of resistors, coils and condensers lead only to rational functions for Z(s) and for T(s). However, the realization of an output like those of Fig. 4., may require for Z(s) (or T(s)) a transcendental function, say F(s). In this case we may have to approximate F(s) by a rational function of s, which, in addition, will have to be also p.r.; this then can be realized by an R-L-C network.

Indeed, once Z(s), or T(s) have been determined, with the restrictions mentioned some very simple rules tell us how to realize the corresponding network, i.e., how to find the geometry of the network and how to compute the numerical values of the resistors, coils, and condensers to be used.

We shall discuss here only how to determine T(s) (the procedure for Z(s) is similar) and ignore the "hardware problem". The interested reader can find exhaustive treatments of this problem, i.e., in [57], or [60]. For the "Bessel case", among others, an almost automatic procedure is outlined in [101]; see also [64].

We start with the consideration of an ideal network that introduces no distortions, only a fixed delay. This means that, given $e_1 = e_1(t)$ for $t \geq 0$, $e_1 = 0$ for $t < 0$, we want to obtain an output $e_2(t) = e_1(t-t_0)$. By taking Laplace transforms we obtain $E_2(s) = \int_0^\infty e_2(t)e^{-st}dt = \int_0^\infty e_1(t-t_o)e^{-st}dt =$

$e^{-st_o}\int_0^\infty e_1(t-t_o)e^{-s(t-t_o)}d(t-t_o) = e^{-st_o}\int_{-t_o}^\infty e_1(t)e^{-st}dt = e^{-st_o}\int_0^\infty e_1(t)e^{-st}dt =$

$e^{-st_o}E_1(s)$.

Let us take, for simplicity, t_o as unit of time (i.e., set $t_o = 1$); we shall indicate later the modifications needed in a different time scale.

We have obtained $E_2(s)/E_1(s) = T(s) = e^{-s}$. As e^{-s} is not rational, one could attempt to approximate it by partial sums of its Maclaurin series. However, not only the partial sums of e^{-s} ($1-s$, $1-s+s^2/2$ etc, have obviously zeros in the right hand plane), but also the partial sums of e^s with more than 4 terms have zeros with positive real parts (see [37]), hence, are not Hurwitz polynomials. Such a transfer function cannot be realized by an R-L-C network. On the other hand, we know from Chapter 10, that $\theta_n(z)$ is a Hurwitz polynomial and from Chapter 13 that $a_n^{-1}\theta_n(s)$ ($a_n = (2n)!/2^n n!$, as in (2.8)) approximates e^s. This leads us to consider a succession of transfer functions $T_n(s)$, that approximate e^s:

(20) $$T_n(s) = a_n\theta_n^{-1}(s) = a_n(\theta_n(s)e^{-s})^{-1}e^{-s} = A_n(s)\cdot e^{-s}.$$

Formula (20) puts into evidence the "distortion factor" $A_n(s) = a_n\theta_n^{-1}(s)e^s$. This approach seems to have been discovered by W.E. Thomson around 1949 (see [107] and [108]), by the study of certain multistage amplifiers that gave desirable outputs. The corresponding transfer functions led him to the BP (at that time not yet so named) and, specifically, to the recursion relation (3.5). Implicit in his work seem to be also Burchnall's relations (10.2), (10.2'). Thomson also tabulates the zeros of $\theta_n(s)$ for $n = 1,2,\dots,9$ to four decimal places (Table 1 in [108]).

If we had $A(s) = 1$, then $T(s) = e^{-s}$ would be the ideal transfer function, that reproduces the input function faithfully at the output with only the fixed delay 1 (or t_o in the general case). In fact, $A_n(s) \neq 1$; specifically, for $s = i\omega$, or rather $s = i\omega t_o$ if $t_o \neq 1$, $A_n(i\omega t_o) = a_n\theta_n^{-1}(i\omega t_o)e^{i\omega t_o} = A_n e^{i\varepsilon}$, say, with $A_n = |A_n(i\omega t_o)|$. For sufficiently large n, $|i\omega t_o|$ is negligible with respect to a_n, so that indeed, $A_n(i\omega t_o) \cong e^{i\omega t_o}$, $|A_n| \cong 1$ and $\varepsilon \cong \omega t_o$. In other words, the filter reproduces faithfully at the output the variation of the applied potential $e_1(t)$, with only a fixed time delay t_o.

On the contrary, if ω is so large that $\theta_n(i\omega t_o) \simeq i^n\omega^n t_o^n$, then $A_n \simeq a_n/\omega^n t_o^n$ and decreases to zero as $\omega \to \infty$. In other words, high frequencies (and what here "high" means depends on n) are practically not transmitted, so that the filter can be made to work (by proper choice of n and by the realization of the network corresponding to that $T_n(s)$) as a low-pass filter.

What happens for intermediate values of ω? At this point it is perhaps worthwhile to remember that T(s) is the result of a Laplace transform so that it should not be considered as the real value of the ratio $e_2(t)/e_1(t)$. Nevertheless, as seen in the two extrema cases, it does convey much information on the dependence of $e_2(t)$ on $e_1(t)$. With this caveat in mind, we now proceed to determine the exact value of $A_n(s)$.

By (5.1), $\theta_n(s)e^{-s} = \sqrt{\frac{2}{\pi}}\, s^{n+1/2}K_{n+1/2}(s)$. By (9.6.4) and (9.1.4) of [1] we obtain

$$\theta_n(s)e^{-s} = \sqrt{\frac{2}{\pi}}\, s^{n+1/2}(-\tfrac{1}{2}\pi i\cdot i^{-(n+1/2)})H^{(2)}_{n+1/2}(-si)$$

$$= -i^{-(n-1/2)}\sqrt{\frac{\pi}{2}}\, s^{n+1/2}H^{(2)}_{n+1/2}(-si)$$

$$= -i^{-(n-1/2)}\sqrt{\frac{\pi}{2}}\, s^{n+1/2}\frac{i}{\sin(n\pi+\pi/2)}(J_{-(n+1/2)}(-si)-i^{2n+1}J_{n+1/2}(-si)).$$

For $s = i\omega$, in particular,

$$\theta_n(i\omega)e^{-i\omega} = -\sqrt{\frac{\pi}{2}}\,(-1)^n(i\omega)^{n+1/2}i^{-(n-3/2)}(J_{-(n+1/2)}(\omega)+(-1)^n(-i)J_{n+1/2}(\omega))$$

$$= \sqrt{\frac{\pi}{2}}\,\omega^{n+1/2}((-1)^nJ_{-(n+1/2)}(\omega)-iJ_{n+1/2}(\omega))$$

and

$$T_n(i\omega) = \frac{a_n}{\omega^{n+1}\sqrt{\frac{\pi}{2\omega}}\{(-1)^nJ_{-(n+1/2)}(\omega)-iJ_{n+1/2}(\omega)\}}\,e^{-i\omega}, \tag{21}$$

a result already obtained by L. Storch (see [106]). If $t_o \neq 1$, we have to replace everywhere ω by ωt_o.

From (21) immediately follows

$$A_n = \frac{a_n}{\omega^{n+1}\{\frac{\pi}{2\omega}(J^2_{-(n+1/2)}(\omega)+J^2_{n+1/2}(\omega)\}^{1/2}}, \tag{22}$$

also found in [106].

Formulae (21) and (22) permit one to compute the loss (usually in decibels), the phase delay and other characteristics of the network. For these topics we have to refer the reader to [106], [64], [101], [115] and [116].

It is of some interest to observe here that the vanishing of many of the sums of odd powers of zeros of the BP (explicitly quoted in [106] and in [108]) is the fundamental reason for the favorable characteristic of the networks based on BP, that earned it the name of maximally flat delay network.

For whatever interest it may have, the general type of network that realizes a transfer function $T_n(s) = (a_n\theta_n^{-1}(s)e^s)$ is:

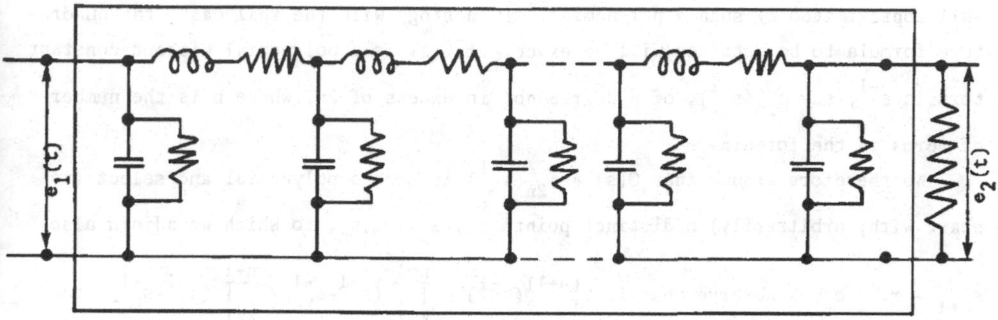

Fig. 5

7. Inversion of Laplace transforms. As already pointed out by Krall and Frink (see, e.g., their remark on p. 106, after (23) in [68]), there seems to exist surprizing analogies between the BP and the Legendre Polynomials. One more instance of such similarity is the following application, that seems to attract increasing attention.

We recall that the Laplace transform $F(s) = \int_0^\infty f(x)e^{-sx}dx$ is inverted by the formula

$$f(x) = \frac{1}{2\pi i} \int_{c-i\infty}^{c+i\infty} F(x)e^{zx}dz \tag{23}$$

valid for Re c sufficiently large.

Often the function F(s) is of a nature that precludes an integration of (23) in closed form. In these cases one is led to numerical integrations.

In the case of a path of integration along the real axis, one favorite method is that of Gaussian quadratures based on n nodes. These formulae use the zeros of the Legendre Polynomials and are exact for polynomials of degrees up to 2n-1. In an analogous way, in the present case, a method of Gaussian quadrature has been devised, with the zeros of the Legendre Polynomials replaced by the zeros of generalized BP. The method is due to H.E. Salzer (see [99] and [100]) and has later been elaborated also by other mathematicians (see [70], [102], [103], [69], [87]). Here we can only sketch the basic ideas and refer the interested reader to the original papers. First, following [99], we simplify (23), by setting zx = s, and obtain

$$xf(x) = \frac{1}{2\pi i} \int_{cx-i\infty}^{cx+i\infty} e^s F(s/x)ds = \frac{1}{2\pi i} \int_{c'-i\infty}^{c'+i\infty} e^s G(s)ds. \tag{24}$$

Here $G(s)$ depends, of course, also on x, but, for simplicity this dependence will not be emphasized by the notation. Also, we shall write again c, rather than c' for the (rather arbitrary) abscissa of integration and denote an integral taken along a parallel to the imaginary axis, at the abscissa c, simply by $\int_{(c)}$. Next, we recall (see, e.g. [13], p.128) that $F(z)$, being a Laplace transform, has to vanish for $z \to \infty$ and satisfies also some other conditions. In many cases of interest it is either a polynomial in z^{-1}, without constant term, or at least can be well approximated by such a polynomial. In analogy with the real case, the quadrature formula to be obtained will be exact, if $G(s)$ is a polynomial without constant term in s^{-1}, say $\rho_{2n}(s^{-1})$, of a degree not in excess of $2n$, where n is the number of terms in the formula.

We therefore assume that $G(s) = \rho_{2n}(s^{-1})$ is such a polynomial and select (to start with, arbitrarily) n distinct points $s_1, s_2, \ldots, s_n$, to which we adjoin also $s_{n+1} = \infty$. We now observe that if $L_i^{(n+1)}(s^{-1}) = \prod_{\substack{k=1\\k\neq i}}^{n+1} (s^{-1}-s_k^{-1}) / \prod_{\substack{k=1\\k\neq i}}^{n+1} (s_i^{-1}-s_k^{-1})$, then $L_i^{(n+1)}(s_k^{-1}) = \delta_{ik}$ (Kronecker delta), so that the Lagrange interpolation polynomial $L^{(n+1)}(s^{-1})$ of degree $n+1$, that coincides with $\rho_{2n}(s^{-1})$ at all points s_k (also at $s_{n+1} = \infty$, where both vanish) is

$$L^{(n+1)}(s^{-1}) = \sum_{i=1}^{n+1} L_i^{(n+1)}(s^{-1})\rho_{2n}(s_i^{-1}).$$

It follows that the polynomial $\rho_{2n}(s^{-1})-L^{(n+1)}(s^{-1})$, of degree $2n$, contains the factor $s^{-1}p_n(s^{-1})$, where $p_n(s^{-1}) = \prod_{k=1}^{n} (s^{-1}-s_k^{-1})$. Hence, $\rho_{2n}(s^{-1}) = L^{(n+1)}(s^{-1}) + s^{-1}p_n(s^{-1})r_{n-1}(s^{-1})$, with a polynomial $r_{n-1}(s^{-1})$ of degree at most $n-1$. Consequently

$$\frac{1}{2\pi i}\int_{(c)} e^s \rho_{2n}(s^{-1})ds = \frac{1}{2\pi i}\int_{(c)} e^s\{L^{(n+1)}(s^{-1})+s^{-1}p_n(s^{-1})r_{n-1}(s^{-1})\}ds =$$

$$\frac{1}{2\pi i}\int_c e^s \sum_{i=1}^{n} L_i^{(n+1)}(s^{-1})\rho_{2n}(s_i^{-1})ds + \frac{1}{2\pi i}\int_{(c)} e^s s^{-1}p_n(s^{-1})r_{n-1}(s^{-1})ds.$$ In the first integral, the last term (corresponding to $s_{n+1} = \infty$) has been left out, because it vanishes on account of $\rho_{2n}(s_{n+1}^{-1}) = \rho_{2n}(0) = 0$. The second integral vanishes identically, if the polynomials $p_n(s^{-1})$ (that is, if the $s_k = s_k^{(n)}$) have been selected in such a way, that the $p_n(s^{-1})$'s are orthogonal to each other, with the weight function $e^s s^{-1}$. Indeed, if

(25) $$\int_{(c)} e^{s} s^{-1} p_n(s^{-1}) p_m(s^{-1}) ds = 0$$

for all $0 \le m < n$, then it is immediate that also

(26) $$\int_{(c)} e^{s} s^{-1} p_n(s^{-1}) s^{-j} ds = 0$$

for $j = 0,1,2,\dots,n-1$, whence

(27) $$\frac{1}{2\pi i} \int_{(c)} e^{s} s^{-1} p_n(s^{-1}) r_{n-1}(s^{-1}) ds = 0$$

follows.

We shall see that such a choice is possible. Assuming this for a moment , (24) becomes

(28) $$x\, f(x) = \sum_{i=1}^{n} A_i^{(n)} \rho_{2n}(s_i^{-1}), \text{ with } A_i^{(n)} = \frac{1}{2\pi i} \int_{(c)} e^{s} s^{-1} L_i^{(n+1)}(s^{-1}) ds.$$

It is clear that the $A_i^{(n)}$, the so called Christoffel constants, are independent of the functions involved, as the $L_i^{(n+1)}$ depend only on the choice of the $s_k^{(n)}$ $(1 \le k \le n)$. The coefficients of $\rho_{2n}(s^{-1})$ depend, of course, also on x.

The Gaussian quadrature formula (28), exact for polynomials $\rho_{2n}(s^{-1})$ without constant term and of degree at most 2n will have been established, as soon as we show how one can select the constants $s_k = s_k^{(n)}$ so that (25) should hold.

In general, $p_n(z) = z^n + \sum_{\nu=1}^{n-1} b_\nu z^\nu$, so that (26) becomes

(29) $$\sum_{\nu=1}^{n} b_\nu \cdot \frac{1}{2\pi i} \int_{(c)} e^{s} s^{-1} \cdot s^{-\nu-j} ds = 0,\ b_n = 1.$$

The integral in (29) equals $\sum_{m=0}^{\infty} \frac{1}{m!} \frac{1}{2\pi i} \int_{(c)} \frac{s^m}{s^{\nu+j+1}} ds$. Here all integrals vanish, except that for $m = \nu+j$, when $\frac{1}{2\pi i} \int_{(c)} \frac{ds}{s} = 1$. The sum over m becomes $\frac{1}{(\nu+j)!}$ and (29) yields

(30) $$\sum_{\nu=0}^{n} \frac{b_\nu}{(\nu+j)!} = 0 \quad \text{for } j = 0,1,\dots,n-1.$$

One may verify that the determinant of the b_ν's $(0 \le \nu \le n-1)$ is different from zero for all $0 \le n \in \mathbf{Z}$, so that the b_ν's $(b_\nu = b_\nu^{(n)})$ are uniquely determined. Once we know that the b_ν's (hence, the $p_n(z)$ and, therefore, the s_k's) are uniquely determined, we can avoid the labor of actually solving (30), by using some of the results of Chapter 4.

We first observe that the infinite vertical path of integration along the abscissa c may be replaced by the open polygonal contour $(-\infty-iY,\ c-iY,\ c+iY,\ -\infty+iY)$

(as $Y \to \infty$, the contributions to the integral in the original and in the modified path, essentially reduce to that from c-iY to c+iY); next this open contour may be replaced by the closed rectangle of vertices $c \pm iY$, $-X \pm iY$ (observe that the integrand has no singularities with Re $s < -X$, if X is sufficiently large); in a third deformation, the rectangle may be replaced by a circle of diameter from -X to c; finally, we invert this circle in the unit circle and take as new variable $z = 1/s$. In this way, it is seen that (25) is equivalent to

$$(31) \qquad \int_C e^{1/z} z^{-1} p_n(z) p_m(z) dz = 0$$

for $m = 0,1,\ldots,n-1$, with C a circle of diameter $(-X^{-1}, c^{-1})$. The only singularity of the integrand is at the origin so that one may actually take for C the unit circle

We now recall Corollary 4.5. First, it is clear that for any a, one may add any constant to $\rho(z;a,b)$, without affecting the validity of the corollary, because the product of two polynomials is holomorphic in the unit circle. We, therefore, may replace in Corollary 5 the weight function $\rho(z;a,b)$, by $\tilde{\rho}(z;a,b) = \sum_{n=1}^{\infty} \frac{\Gamma(a)}{\Gamma(a+n-1)} (-\frac{b}{z})^n$. Next, we consider the particular case $a = 1$, $b = -1$, so that $\tilde{\rho}(z;1,-1) = \sum_{n=1}^{\infty} (\Gamma(n))^{-1} z^{-n} = z^{-1} e^{1/z}$. Corollary 5 now reads

$$\frac{1}{2\pi i} \int_{|z|=1} e^{1/z} y_n(z;1,-1) y_m(z;1,-1) \frac{dz}{z} = 0$$

for $0 \leq m < n$.

It follows that if we take in (31) $p_n(z) = c_n y_n(z;1,-1)$, then (25), hence, also (26) and (27) hold. Here $c_n = \{(-1)^n n(n+1)\ldots(2n-1)\}^{-1}$, as follows by comparing the coefficients of z^n on both sides of the equality. On the other hand, it follows from the uniqueness of the b_ν's (i.e., of the $p_n(s)$'s), that the coefficients of $c_n y_n(z;1,-1)$ are the only solution of (30). The $s_k^{(n)}$ $(k = 1,2,\ldots,n)$ are then the reciprocals of the zeros of $y_n(z;1,-1)$, i.e., they are the zeros of the polynomials $\theta_n(z;1,-1)$. This finishes the proof of the exact quadrature formula (28).

In case $G(s^{-1})$ is not a polynomial of degree $\leq 2n$, then $G(s^{-1}) = L^{(n+1)}(s^{-1}) + s^{-1} p_n(s^{-1}) g(s^{-1})$, where g(z) is not, in general, a polynomial of degree $\leq n-1$ but is holomorphic for Re $s > c$ (see [13]). Instead of (28) we now obtain

$$xf(x) = \sum_{i=1}^{n} A_i^{(n)} G(s_i^{-1}) + R,$$

where $R = \frac{1}{2\pi i} \int_{(c)} e^s s^{-1} c_n y_n(s^{-1}) g(s^{-1}) ds$ is an error term. While this, in general, is not identically zero, it may still be possible to obtain an upper bound for

$|R|$, in a way similar to the classical case of real paths of integration and (28), while not exact, gives a usable approximation to the inversion.

In [99] Salzer lists the polynomials $y_n(z;1,-1)$ for $n = 1(1)12$ (this notation means that n runs from 1 to 12, with the difference between consecutive entries indicated in the parentheses), the zeros of the corresponding $\theta_n(z;1,-1)$ to 8 significant figures and coefficients $A_n^{(i)}$ to 7 or 8 figures.

In [100] Salzer completes the theory of BP, by quoting Krall and Frink, Burchnall, Agarwal, Al Salam, Ragab and Brafman. He extends his list of polynomials to $n = 16$, and tabulates the zeros of the $\theta_n(z;1,-1)$ and the Christoffel numbers for $n = 1(1)16$ to 15 significant figures.

Already in 1959, Kublanowskaya and Smirnova [70] had published a tabulation of the zeros of $\theta_n(z)$ $(= \theta_n(z;2,2))$ and of $n\theta_n(z) + z^2\theta_{n-1}(z)$ for $n = 1(1)30$. Other tables for the numerical inversion of the Laplace transform were published by Skoblja [102] for s and A_k in the formula $\frac{1}{2\pi i}\int_{(c)} e^z z^{-s}\phi(z)dz \cong \sum_{k=1}^{n} A_k^{(n)}\phi(z_k)$ with $n = 1(1)10$ and $s = .1\ (.1)3$, to about 8 and 7 significant figures for the z_k and $A_k^{(n)}$, respectively. See also [103]. Next, Krylov and Skoblja, in [69] tabulate s and A_k, as follows: $s = 1(1)5$ for n 1(1)15 to 20 significant digits and for $s = .01\ (.01)3$ and $n = 1(1)10$ to 7 or 8 significant digits. A generalization of these procedures can be found in R. Piessens [87]. Further comments may be found in the books of Y.L. Luke [71], vol. II, p. 189 et seg., and [72], p. 229 and 433.

CHAPTER 15

MISCELLANEA

Many papers related to BP have not been considered in detail in the preceding chapters. Sometimes, when such a work was related to a topic discussed, or was a rediscovery of an older result, it was at least quoted. This was the case, e.g., with many papers on generating functions of BP. In other cases, however, when a paper treated some isolated topic, not easily fitted into some broader category, the paper often was not mentioned at all. Among these papers are some very valuable ones, other of a more routine character and also some (fortunately very few) with erroneous claims. It also is not surprizing that among the so far ignored papers, many are by authors some of whose other work on BP has already been prominently displayed.

The purpose of the present chapter is to attempt to mention all these, so far neglected papers and their authors and to state at least briefly and without proofs the results obtained.

A major difficulty is to make such a presentation in a systematic way. It could, perhaps, be considered desirable to arrange the material by subject matter. In fact, when several papers discussed related topics, in general they could already be fitted into one of the preceding chapters. Most of the material here considered, consists of papers each of which discusses a more or less isolated subject.

In spite of many advantages, also a strictly chronological presentation would not be very useful. Indeed, some authors worked on different topics related to BP, throughout many years and their names would be intermingled in a rather confusing way. For these reasons it has appeared that the most convenient way is to list these papers by authors. In order to facilitate the search for a specific paper, whose author is known, the authors are listed in alphabetic order. The papers of any given author, however, will then be listed in chronological order, under the heading of their author.

With very few exceptions, it has been possible to obtain the original papers. Whenever this turned out to be impossible, or to require an excessive investment of time, the information concerning its contents was obtained through the reviewing journals, especially the Mathematical Reviews.

Even in the cases, when the original paper was accessible, no critical appraisal of its contents appeared possible, or even desirable. An exception was made when a result appeared to be plainly erroneous, or valid only under restrictions not stated explicitly.

1. W.H. Abdi [1]. We recall the definition of the basic hypergeometric series (see, e.g. [6]). A new parameter q is introduced, $|q| < 1$. The ratios $\frac{a(a+1)\ldots(a+n-1)}{c(c+1)\ldots(c+n-1)}$ that occur in the definition of the hypergeometric series, are replaced by

$\frac{(1-q^a)(1-q^{a+1})\dots(a-q^{a+n-1})}{(1-q^c)(1-q^{c+1})\dots(1-q^{c+n-1})}$. One observes that the limit of the new ratio, for $q \to 1$, is the previous ratio. In analogy to the definition of the basic hypergeometric function, the author defines a basic analog of BP. For this he starts from ${}_2\phi_0(q^{-n},q^{a+n-1};-,x)$ (the basic analog of ${}_2F_0(-n,a+n-1;-;-x/b)$, (see [6]), but with x instead of -x/b) and defines the basic BP by

(1) $$J(q;a-1;n;x) = \{(1-q^{a-1})_n/(q;n)\}\ {}_2\phi_0(q^{-n};q^{a+n-1};x), \text{ where}$$

$$(1+x)_n = \prod_{k=0}^{\infty} (1+xq^k)/(1+xq^{k+n}), \text{ and } (1-q)_n = (q;n).$$

The author studies many properties of $J(q;c,n;x)$, such as integral representations, recurrence relations, etc.

The motivation for this particular definition (1) may, presumably, be found in Rainville's [90] normalization $\phi_n(a-1,x) = \frac{(a-1)_n}{n!}\ {}_2F_0(-n,a+n-1;-;x)$. On the one hand, this coincides with $\frac{(a-1)_n}{n!}\ y_n(-bx;a,b)$; on the other hand, in the spirit of a "basic" function, $\lim_{q \to 1^-} J(q;a-1;n;x) = \phi_n(a-1,x)$. Otherwise, the connection of $J(q;a-1;n,x)$ to $y_n(x;a,b)$ is rather tenuous.

2. R.R. Agarwal. In [2], in addition to the material mentioned earlier (see Chapter 5), one finds also many other results, such as:

(a) Some additional recurrence relations for $y_n(z;a,b)$.

(b) Integral representations, like

$$y_n(z;a,b) = \frac{n!}{2\pi i}\left(\frac{z}{b}\right)^n \int_{|u|=1} \frac{u^{2-a}}{u(1-u)^{n+1}} e^{bu/z} du,$$

and

$$y_n(z;a,b) = \frac{1}{\Gamma(n+a-1)} \int_0^\infty t^{a-2+n}(1+tx/b)^n e^{-t} dt.$$

(c) Expansion formulae, of which the following is typical:

$$y_{2n}(2x;a,b) = \sum_{r=0}^{2n} \frac{(-2n)_{2r}(a+2n-1)_r}{r!} \left(\frac{x}{b}\right)^{2r} y_{2n-2r}(x;2a+2n+4r-1,b).$$

3. W.A. Al-Salam. In addition to the contributions already presented, the following should be mentioned.

In [3] we find:

(a) New integral representations and new generating functions.

(b) A connection with Jacobi Polynomials:

$$y_n(x;a) = n!(-x/2)^n \lim_{\beta\to\infty} P_n^{(-2n-a+1,\beta)}(1-(x\beta)^{-1}).$$

(c) Characterizations of BP, as follows:

(i) Let a sequence of polynomials $f_n(x,a)$, of degrees $n = 1,2,\ldots$, depending also on a parameter a, and normalized by $f_n(0,a) = 1$, satisfy the recurrence relation (due to Krall and Frink; see Section 19 of the present chapter)

$$f'_n(x,a-2) = \frac{1}{2} n(n+a-1) f_{n-1}(x,a); \tag{2}$$

then $f_n(x,a) = y_n(x;a)$.

(ii) Let $\Delta_a y_n(x,a) = y_n(x;a+1) - y_n(x;a)$; then

$$\Delta_a y_n(x;a) = \frac{1}{2} nxy_{n-1}(x,a+2),$$

and, more generally,

$$\Delta_a^k y_n(x;a) = n^{(k)}(x/2)^k y_{n-k}(x;a+2k);$$

in particular,

$$\Delta_a^n y_n(x;a) = n!(x/2)^n.$$

From this and Newton's formula $f(\alpha+\mu) = \sum_r \binom{\mu}{r} \Delta^r f(\alpha)$ the author infers that the BP satisfy the identity

$$\Delta_a f_n(x,a) = \{x/(n+a-1)\} f'(x,a). \tag{3}$$

Let now $f_n(x,a)$ be any sequence of functions, not necessarily polynomials, depending, besides the variable x, also on a parameter a. Furthermore, let $f_n(x,a)$ be normalized by $f_n(0,a) = f_o(x,a) = 1$ and assume that the $f_n(x,a)$ satisfy (3) and the "initial condition" $f_n(x,2) = y_n(x)$; then $f_n(x,a) = y_n(x;a)$.

(d) Expansion formulae, such as

$$y_n(x) = 1 + x\{(2n-1)y_{n-1}(x) + (2n-3)y_{n-3}(x) + \ldots\};$$

$$\sum_{k=o}^{\infty} \frac{(-\lambda)^k}{k!} y_n(x,k-n+2) = e^{-\lambda} \sum_{s=o}^{n} (-n)_s (-x/2)^s L_s^{(o)}(\lambda) \quad (L_s^{(o)}(\lambda) = \text{Laguerre Polynomial});$$

$$\sum_{k=o}^{\infty} (-1)^k \lambda^{2k} y_n(x,2k-n+2) = \frac{1}{2\lambda} \int_o^\infty e^{-t}(1+xt/2)^n \frac{\sin 2\lambda t}{t}\, dt;$$

$$(z/2)^{\beta-1} e^{xz^2/8} = \sum_{n=o}^{\infty} (2n+\beta-1) \frac{\Gamma(n+\beta-1)}{n!} y_n(x;\beta) J_{2n+\beta-1}(z)$$

($\beta \neq 0, -1, -2, \ldots$; $J_\alpha(z)$ = Bessel function).

(e) Product expansions (see also Brafman [15]):

$$y_n(u,a)y_n(v,a) = \sum_{r=0}^{n} \frac{(-n)_r(n+a-1)_r}{r!} \left(-\frac{u+v}{2}\right)^2 y_r\left(\frac{uv}{u+v},a\right)$$

and the corresponding inversions

$$\left(-\frac{u+v}{2}\right)^n y_n\left(\frac{uv}{u+v},a\right) = \sum_{s=0}^{n} \frac{(-n)_s(2s+a-1)\Gamma(s+a-1)}{s!\Gamma(n+s+a)} y_s(u,a)y_s(v,a);$$

also similar formulae involving $\theta_n(x,a)$.

(f) Product representation formulae, like the following two:

$$y_n(u,a)y_n(v,a) = \frac{n!}{\Gamma(n+a-1)} \int_0^\infty t^{a-2} P_n^{(a-2,0)}\left(1+t\left(u+v+\frac{uvt}{2}\right)\right)e^{-t}dt$$

and

$$y_n\left(\frac{x}{s}\right)y_n\left(-\frac{x}{s}\right) = \frac{1}{2}\left(\frac{x}{s}\right)^{2n} \int_0^\infty t^{2n} L_n^{(-n-1/2)}(-s^2/2tx^2)e^{-t/2}dt$$

$$= \frac{s^{2n+2}}{n!} \int_0^\infty t^n y_n\left(-8x^2t, \frac{3}{2}-n\right)e^{-ts^2}dt.$$

In [4] the author considers the second (non-polynomial) solution of (2.11), which is $q_n(x) = (-1)^n e^{2/x} y_n(x)$, and establishes recurrences for $q_n(x)$, analogous to (3.16) and (3.22) (for $a = 2$). He also gives new characterizations of the BP, of which the following is typical:

Given a sequence of polynomials $f_n(x)$, of degrees n ($n = 0,1,2,\ldots$), normalized by $f_0(x) = 1$, then $f_n(x) = y_n(x)$ if and only if

$$x^2 \frac{d}{dx}(f_n(x)f_{n-1}(x)) = (f_n(x)-f_{n-1}(x))^2. \tag{4}$$

In [5] the author obtains a second solution $z_n^{(a)}(x)$ to (2.26) and a second polynomial solution $V_n^{(a)}(x)$ to (3.20). $z_n^{(a)}(x)$ can be represented by a definite integral and also in terms of Whittaker functions. For integral $a \geq -n$, $z_n^{(a)}(x)$ is an entire function; otherwise it is holomorphic only in the plane cut along the positive real axis, where $z_n^{(a)}(x)$ has along $x > 0$, the discontinuity $-2i\sin(a\pi)y_n(x;a)$. For $z_n^{(a)}(x)$, the author proves recurrence relations, generating functions, multiplication theorems, etc.

The "associate polynomials" $V_n^{(a)}(x)$ are orthogonal on the unit circle with respect to the weight function $\Omega(x,a) = x\{{}_1F_1(1;a;-zx^{-1})\}^{-1} = \sum_{r=0}^{\infty} \beta_r(a)x^{1-r}$; here the coefficients may be computed recursively by $\sum_{r=0}^{m} (-2)^r \beta_{m-r}(a)/(a+1) = \beta_m(a)$ for

$m \geq 0$, $\beta_o(a) = 1$; for $a = 2$, $\beta_n(2) = (-1)^n 2^n B_n/n!$, with B_n the Bernoulli numbers. As particular cases, the author recovers several results of Al-Salam and Carlitz [7]. Next, some very general identities of Christoffel-Darboux type are presented. Several properties of the BP $y_n(z;a)$ are obtained, some of them new, others already known (but often with new proofs), such as orthogonality (see [68] and [93]), simplicity of zeros (see [53]), and others. The author shows that, for $a > 2$, all zeros of $y_n(z;a)$ are inside the unit circle (for $a = 2$ the result had been proved in [53]) and those of $V_n^{(a)}(x)$ inside $|x| \leq \sqrt{3/5}$. Also the irreducibility of $y_n(z;a)$, for a an integer, is proved for certain values of n.

In [6] the author generalizes some results of Chatterjea [23].

4. W.A. Al-Salam and L. Carlitz. In [7] the authors study polynomials $u_n(z)$, $v_n(z)$, which are, essentially, the polynomials $P_n(z)$, $Q_n(z)$ of Chapter 8. Here are some characteristic results:

Define coefficients β_n and γ_n by $\beta_n = 2^n B_n/n!$ (B_n = Bernoulli numbers) and $2(e^x+1)^{-1} = \sum_{n=0}^{\infty} \gamma_n (x/2)^n$; then, for $0 \leq m \leq n$, $\gamma^{m+1} u_n(\gamma) = (-1)^n \frac{2^n n!}{(2n+1)!} \delta_{mn}$ and $\beta^{m+2} v_n(\beta) = (-1)^n \frac{2^{n+1}(n+1)!}{(2n+3)!} \delta_{mn}$. Here the δ's are Kronecker deltas and the "powers" are symbolic: after $\gamma^{m+1}u_n(\gamma)$ and $\beta^{m+2}v_n(\beta)$ are computed, the exponents are lowered to become indeces. Also,

$$-\frac{1}{2\pi i}\int_{|z|=1} \frac{2}{e^{2/z}-1} z^m u_n(z)dz = (2n+3)(2n+2)\cdot\frac{1}{2\pi i}\int_{|z|=1}\frac{2}{e^{2/z}-1} z^m v_n(z) = \frac{(-1)^{n+1}(n+1)!2^{n+1}}{(2n+1)!}\delta_{mn}.$$

In [8] the authors study the related polynomials $U_n(x) = i^{-n}u_n(ix)$, $V_n(x) = i^{-n}v_n(ix)$. Let $\alpha(x)$ be a step function with salti $j(x_k) = x_k^2$ at $x_k = \frac{2}{\pi}\frac{1}{2k+1}$ ($k \in \mathbf{Z}$); then

$$\int_{-\infty}^{\infty} x^m U_n(x)d\alpha(x) = \frac{2^n n!}{(2n+1)!}\delta_{mn}, \text{ and } \int_{-\infty}^{\infty} U_m(x)U_n(x)d\alpha(x) = \frac{\delta_{mn}}{2n+1}.$$

Similarly, if $\beta(x)$ is a step function with salti $j(x_k') = \frac{3}{\pi^2 k^2}$ at the points $x_k' = \frac{1}{2\pi k}$ ($k = \pm 1, \pm 2, \ldots$), then

$$\int_{-\infty}^{\infty} V_m(x)V_n(x)d\alpha(x) = \frac{3}{2n+3}\delta_{mn}.$$

5. W.A. Al-Salam and T.S. Chihara. The classical polynomials of Jacobi, Hermite, and Laguerre are the only orthogonal polynomial solutions of

(5) $$\pi(x)P_n'(x) = (\alpha_n x+\beta_n)P_n(x) + \gamma_n P_{n-1}(x) \quad (n \geq 1),$$

with $\pi(x)$ a polynomial and α_n, β_n, γ_n constants.

In [9] the authors show that, if the classical meaning of orthogonality is relaxed to admit weight functions of bounded variations, then the BP have to be added to the solutions of (5) (indeed, see (3.10)). This result solves affirmatively an older conjecture of Karlin and Szegö.

6. I. Baičev [10] studies the convergence and, especially the C-1 summability of series of BP.

7. D.P. Banerjee [11] discusses some non-linear recursions; his "Turán inequality" is incorrect as printed (see [28] for correct form; an equivalent result had been obtained earlier by L. Carlitz [19]).

8. L. Carlitz defines (see [19]) the functions $f_n(x) = x\theta_{n-1}(x)$ introduced in Section 6.7. In addition to the results discussed there, he proves recursion identities for $f_n(x)$, the formula

$$f_n(u+v) = \sum_{r=0}^{n} \binom{n}{r} f_r(u) f_{n-r}(v);$$

also representations of Laguerre Polynomials and of BP as sums of $f_n(x)$'s. He shows that

$$\theta_n(x)\theta_n(y) = \sum_{r=0}^{n} \frac{(n+r)!}{(n-r)!r!2^r}(xy)^{n-r}\theta_r(x+y),$$

and proves the inequalities of Turán type $\theta_{n+1}\theta_{n-1}-\theta_n^2 \geq 0$ and $\theta_n\theta_{n+1}'-\theta_{n+1}\theta_n' \geq 0$ for $x \geq 0$. Finally, the author investigates some arithmetic properties of $\theta_n(z)$.

9. S.K. Chatterjea published at least 15 papers related to BP. In addition to the author's work quoted in Chapter 6, the following deserves mention. In [21] he gives two representations of $y_n(z)$ as a determinant (one of them essentially equivalent to that discussed in Chapter 3). In [22] he proves the formula $y_{2n}y_{2n+2} = y_{2n}^2 + (4n+3)x^2 \sum_{k=1}^{2n} (2k+1)y_k^2 + (4n+3)x(x+1)$. From this it easily follows that $y_{2n}(x) > 0$ for $x \leq -1$, a result used by Cima [40] in his proof that $y_{2n}(x) \neq 0$ for all real x.

In a sequence of three papers [23], [24], [25], the author establishes and then uses operational formulae. In [23] he proves (with $D = \frac{d}{dx}$) that $\prod_{j=1}^{n} \{x^2D+(2j+a)x+b\} = \sum_{r=0}^{n} b^{n-r}x^{2r}y_{n-r}(x;a+2r+2,b)D^r$, which he uses to show that

$$y_{n+m}(x;a,b) = \sum_{r=0}^{\min(n,m)} \binom{m}{r}\binom{n}{r} r!(m+2n+a-1)_r\left(\frac{x}{b}\right)^{2r} y_{n-r}(x;a+2r,b)y_{m-r}(x;a+2n+2r,b).$$

In [24] he proves the formulae

$$x^{2n}(D+2(nx+1)x^{-2})^n = \sum_{r=o}^{n} \binom{n}{r} 2^{n-r} x^{2r} y_{n-r}(x;2+2r,2)D^r;$$

and

$$x^n(D-(2x+n+1)x^{-1})^n = \sum_{r=o}^{n} (-2)^{n-r} x^r \theta_{n-r}(x;2+r,2)D^r.$$

In [25] the author proves that

$$e^{bx} \prod_{j=1}^{n} (xD-a-2n+j+1)e^{-bx}y = \sum_{r=o}^{n} \binom{n}{r}(-b)^{n-r} x^r \theta_{n-r}(x;a+r,b)D^r y,$$

from which follows that

$$(-b)^n \theta_n(x;a,b) = \prod_{j=1}^{n} (xD-bx-a-2n+j+1)\cdot 1.$$

In [26] the author obtains a representation of the BP as a double integral.

In [27] the author gives a new generalization of the BP.

In [28] some of the Christoffel-Darboux identities contained in (or easily derived from) [5] are rediscovered.

In [29] the author considers the polynomials defined by

$$M_n(x,k) = x^{(2-k)n} k^{-n} e^{k/x} D^n (x^{kn} e^{-k/x}) \quad (2 \le k \in \mathbf{Z}).$$

These polynomials generalize the BP, to which they reduce for $k = 2$ (see (7.2")), while $M_n(x,3) = y_n(x;n+2,3)$.

In [32] the author proves Turán's inequality $\Delta_n(x) = y_n y_{n+2} - y_{n+1}^2 \ge 0$ for $x \ge 0$ (which as already observed is actually an immediate consequence of Carlitz's result in [19]), by first obtaining an explicit formula for $\Delta_n(x)$. He then proceeds to use this formula to prove also

$$\int_{|x|=1} (x^{-2}\Delta_n(x)-x^{-1})e^{-2/x}dx = 8\pi i(-1)^{n+1}(1+[\tfrac{n}{2}]).$$

In [33], in analogy with Burchnall's operator $Q(\delta)$ (see [17] and Theorem 2.2) the author introduces the differential operator $\tilde{Q}(\delta) = \delta(\delta-2)\ldots(\delta-2n+2)$ and studies the polynomials defined by $\phi_n(z) = (-1)^n e^z \tilde{Q}(\delta)e^{-z}$ $(n \ge 1)$, $\phi_o(z) = 1$.

In [34] the author gives a proof, different from Carlitz's [19], for Turán's inequality

$$\theta_{n-1}(x)\theta_{n+1}(x)-\theta_n^2(x) \ge 0 \text{ for } n \ge 1,\ |x| \le 1.$$

10. <u>C.K. Chatterjee</u> [38] proves (a) that for $n > 1$, $y_n'(x)$ and $y_{n-1}'(x)$ have no common zero; (b) the relation $(x^3 \frac{d}{dx})^k (x^n e^{-1/x} y_{n-1}(x)) = x^{n+k} e^{-1/x} y_{n+k-1}(x)$ and similar ones; and (c) some inequalities for products of BP.

11. J.A. Cima [40] gives a correct proof for the statement that $y_{2n}(x) \neq 0$ for all real x, whose proof in [53] had been incorrect.

12. M.K. Das published at least 5 papers related to BP. In addition to his work mentioned in Chapter 6, we find in [41] that, if Y is a sufficiently differentiably function and $\delta = x \frac{d}{dx}$, as in Chapter 2, then

$$[x(\delta+a+k-1)]^n(x^{n+k}e^{-b/x}Y) = e^{-b/x} \sum_{p=0}^{n} \binom{n}{p} b^{n-p} x^{n+2p+k} y_{n-p}(x;a+2p+2,b) D^p Y,$$

and the author makes some nice applications of this and related formulae.

In [45] the author proves a very general operational identity. As particular cases he obtains, among others, again the formula of [41].

13. D. Dickinson's paper [47] contains, in addition to the work discussed in Chapters 5, 9, and 10, and material not directly related to BP, also the following result:

Let D be an arbitrary finite set of integers and let s be a fixed integer. Also, let $f_n(x)$ be a sequence of algebraic functions such that

$\sum_{n \in D} f_n(x) J_{n+s+1/2}(s)$ vanishes identially ($J_\alpha(s)$ = Bessel function); then

both sums, $\sum_{n \in D} f_n(1/ix) i^n y_{n+s}(x)$ and $\sum_{n \in D} f_n(-1/ix) i^{-n} y_{n+s}(x)$ also vanish identically (observe the misprint s for x in the theorem of [47]).

14. K. Dočev [48] completes and refines work of Obreshkov [82], both, on the zeros of BP and on the expansions of functions in series of BP. The author's contribution to the first topic was presented in Chapter 10. In connection with the second, the author proves the following Abelian theorem: Let us assume that the series

$\sum_{n=0}^{\infty} a_n y_n(x;a,-1)$ converges in a subset D of $|z| < |x_0|$ to the function f(x), that

x_0 belongs to the closure of D, and that the series converges also at $x = x_0$ to s;

then $\lim_{x \to x_0} f(x) = s$, provided that the path along which x approaches x_0 is inside D.

15. M. Dutta S.K. Chatterjea, M.L. Moore study in [49] a class of orthogonal polynomials, that can be represented by $H_n^\mu(x) = x^n \, {}_2F_0(-\frac{n}{2}, -\frac{1}{2}(n+\mu-1);-;-1/x^2)$ and note the similarity of these polynomials to the BP.

16. M.T. Eweida [51] indicates connections of BP with Meijer's G-functions, Laguerre Polynomials and Bessel Functions and gives integral representations, of which the following are characteristic:

$$\theta_n(x) = \pi^{-1}2^{n+1}n!e^x e^{2n+1} \int_o^\infty \frac{\cos t}{(t^2+n^2)^n}\,dt$$

$$\theta_n(x;a) = (-1)^{n+a-1}2^{(a-1)/2}e^{2x}x^{(2n+a-1)/2} \int_o^\infty e^{-t}t^{-(a-1)/2}J_{2n+a-1}(2\sqrt{2tx})dt$$

$$\theta_n(u)\theta_n(v) = (2\pi)^{-1/2}e^{u+v} \int_o^\infty c^{-(t+(u+v)^2/t)/2}t^{n-1/2}\theta_n(\tfrac{u+v}{t})dt.$$

In [52] the author rephrases some (known) infinite integrals that involve $K_{n+1/2}(x)$, by replacing the Bessel function by its value (see Section 3.1) $2^{-1/2}\pi^{1/2}e^{-x}x^{-n-1/2}\theta_n(x)$.

17. A.M. Hamza [59] evaluates some integrals that involve BP.

18. M.E.H. Ismail [61] generalizes and solves a problem that had previously been asked by R. Askey [2]. The problem , in its generalized formulation, is the following: Let $P_n(x,a)$ be a sequence of polynomials, orthogonal over an interval I, with respect to a weight function $\omega(x,a)$, where both, the polynomials and the weight function depend besides the variable x also on a parameter a. Given also a sequence a_n $(n = 0,1,2,\ldots)$, integers N and M, and constants a,b,c,d, it is required to determine a function f(x) in such a way that

$$a_n = \begin{cases} \int_I f(x)\omega(x,c)P_n(x,a)dx & \text{for } n = 0,1,\ldots,N; \\ \int_I f(x)\omega(x,c)P_n(x,d)dx & \text{for } n = N+1,\ N+2,\ldots,M; \\ \int_I f(x)\omega(x,b)P_n(x,b)dx & \text{for } n \geq M+1. \end{cases}$$

Conditions on $P_n(x,a)$, necessary for the solvability of this problem are determined and among the solutions $P_n(x,a)$ are found also the BP.

The author obtains as one of his results also the following formula (previously obtained by Al-Salam [3]):

$$y_n(x;a) = \sum_{k=o}^{\infty} \binom{n}{k} \frac{(a+n-1)_k}{(c+n-1)_k} \frac{(4k+c-n-a)_{n-k}}{(2k+c)_{n-k}} y_k(x;c).$$

Other contributions of the author with Kelker to the theory and applications of BP have been mentioned in the Introduction, and in Chapters 1, 10, and 14.

19. H.L. Krall and O. Frink coined the name Bessel Polynomials in [68] and much of their work has been discussed in preceding chapters. Nevertheless [68], contains also some topics, mainly related to the generalized BP $y_n(z;a,b)$, $a \neq 2$, that were not discussed so far and they deserve mention at this place.

Several recurrence relations occur in [68], not mentioned in Chapter 3. Also the differential equation for $y'(z;a,b)$,

$$x^2 \frac{d^2y_n'}{dx^2} + (ax+2x+b)\frac{dy_n'}{dx} = n(n+a+1)y_n' \tag{6}$$

appears in [68]. From (6) it follows that $y_n'(z;a,b)$ is proportional to a BP with the value a+2 for its principal parameter. As $y_n'(z;a,b)$ is a polynomial of degree n-1, it follows that $y_n'(z;a,b) = c_1y_{n-1}(z;a+2,b)$, or $y_n'(z;a-2,b) = cy_n(z;a,b)$. By comparing the constant terms on both sides we obtain that $c = n(n+a-1)/2$. It is this result that was generalized by Al-Salam [3] to (2).

20. P. J. McCarthy's contribution in [74] to the location of the zeros of BP was mentioned in Chapter 10. In [75] he shows how to generalize Al-Salam's formula (4) (see [4]) to other systems of orthogonal polynomials.

21. N. Obreshkov's paper [82] has already been mentioned in connection with Dočev's work on the zeros of BP in Chapter 10. In the same paper the author also studies expansions of holomorphic functions in series of BP.

22. A. Pham-Ngoc Dinh [86] studies the function $Y_n^a(z) = e^{-z}z^{(1-a-2n)/n}\theta_n(z,a)$. The author obtains differential-difference relations satisfied by $Y_n^a(z)$, operators that raise, or lower the value of the parameter a, and expansions of $Y_n^a(z-utz^{1/2})$ in series of the form $\sum_{p=o}^{\infty} c_n(p,z,tu)Y_n^{a+p}(z)$, with rather complicated coefficients $c_n(p,z,tu)$, and convergent for $|ut| \leq \sqrt{|z|}$, unless a is a negative integer or a = +1.

23. F.M. Ragab [88] states and offers proofs for five summation formulae, of which the following is the simplest:

$$\sum_{r=o}^{n} \binom{n}{r} \frac{\Gamma(2-a)}{\Gamma(2-a-r)} \left(-\frac{2x}{b}\right)^r y_r\left(\frac{x}{2}\,;\,a,b\right) = y_{2n}(x;a-2n,b). \tag{7}$$

It seems that certain restrictions (not explicitly stated in [88]) on the values of a are needed, in order to insure the validity of the summation formulae. So, e.g., if the arguments of the gamma functions are zero, or negative integers, the ratio $\frac{\Gamma(2-a)}{\Gamma(2-a-r)}$ presumably has to be interpreted as its limiting value $(-1)^r(a-1)a(a+1)\dots(a+r-2)$. For a = 1 this product vanishes, except when r = 0, so that (7) becomes $y_o(\frac{x}{2}\,;\,1,b) = y_{2n}(x;1-2n,b)$, which is obviously incorrect. Indeed $y_o(\frac{x}{2}\,;\,1,b) = 1$, while $y_{2n}(x;1-2n,b)$ is a polynomial of degree 2n. For n = 1, e.g., $y_2(x;-1,b) = 1+(4/b)x+(6/b^2)x^2$.

24. A.K. Rajagopal [91] verifies that a certain function, closely related to the BP satisfies a certain equation (known as Truesdall's equation) and from this fact he draws some (essentially known) conclusions about the BP themselves.

25. H. van Rossum, in [93] and [94] studies the orthogonality of the BP on hand of the Padé table. He calls a sequence of real numbers c_m "strictly totally positive" if all determinants $|c_{m+i-j}|$ $(i,j = 0,1,2,\ldots,n-1)$ are positive for all $m = 0,1,2,\ldots$ and $n = 1,2,3,\ldots$. If the Padé rational fraction of the square (μ,ν) of the power series $\sum_{m=0}^{\infty} c_m z^m$ has the denominator $V_{\mu,\nu}(z)$, then the polynomials $B_\mu^{(k)}(z) = z^\mu V_{k,k+\mu}(-z^{-1})$ (k = fixed integer, $\mu = 0,1,\ldots$) are said to be "totally positive". They are orthogonal on the circle $|z| = \rho+\varepsilon$ ($\rho = \limsup \{m\sqrt{c_m}\}$, $\varepsilon > 0$), with respect to the weight function $\phi^{(k)}(z) = \sum_{m=0}^{\infty} (-1)^{m+k+1} c_{m+k+1} z^{-m-1}$; they have only positive coefficients; and they have all their zeros inside the circle $|z| < c_1/c_0$. The BP are an instance of totally positive polynomials, corresponding to $c_0 = c_1 = 1$. The author's further contributions to the location of zeros of BP [95] have already been mentioned in Chapter 10.

26. P. Rusev [96] indicates sufficient conditions for the convergence of series in BP, similar to those corresponding to series in Jacobi polynomials inside an ellipse with foci at +1 and -1.

27. H. Rutishauser [97] uses continued fractions, in order to give elegant proofs of results previously obtained by Al-Salam and Carlitz [8] and by D. Dickinson [47], concerning the orthogonality on the real axis of certain functions (denoted in [8] by $U_n(x)$ and $V_n(x)$), closely related to BP.

28. S. L. Soni [104] obtains operational formulae for the generalized BP by using operators different from $D(= d/dx)$ and $\delta(= x\, d/dx)$.

29. H. M. Srivastava [105] establishes expansions of a large class of functions in series of von Neumann type, involving (among other systems) also BP.

30. L. Toscano, in addition to his work on generating functions [109] quoted in Chapter 6, also obtain (in [110])

(a) representations of BP as Laplace transforms of a Legendre Polynomial: $y_n(-2/p) = p \int_0^\infty L_n(1-2t)e^{-pt}dt$, valid for $\operatorname{Re} p > 0$; and, in a somewhat changed notation,

(b) representations of the type

$$y_n(x^{-1};a) = x^{-n}\theta_n(x;a) = \Gamma(\alpha)^{-1} \int_0^\infty u^{\alpha-1} {}_2F_1(-n,n+a-1;\alpha,-ux/2)e^{-u}du.$$

31. It is not possible to conclude this chapter, as well as this monograph, without the sincerest apologies of the present author to the many contributors to the theory, or the applications of BP, whose work was either overlooked, or misinterpreted, or underevaluated. Ideally, every one of them, who could be located, ought to have received the intended text (at least in as far as it related to his/her work), with a request for approval and/or comments. Due to the large number of contributing mathematicians, their dispersal throughout the world, the many years that have elapsed since many of the papers were written - and, last, but not least, the fact that, unfortunately, some of them are no longer alive, this ideal procedure was not practicable. Even the partial execution of the possible part of this program would have entailed such an additional delay for the present publication, that the value of this monograph itself would have been put in question. For this reason the present author counts on the understanding of his colleagues, and on their forgiveness of his sins of omission and commission.

APPENDIX

SOME OPEN PROBLEMS RELATED TO BP

Quite frequently papers appear with improvements of known results concerning BP. So, e.g., new generating functions are obtained, sharper bounds for the location of the zeros of BP are determined, new recursion formulae are established, etc. Such improvements of known results are of course always welcome; here, however, we want to call attention to a number of unsolved problems related to BP and of a somewhat different nature.

1. It seems that the polynomial solutions of equation (2.26), with $a = 1$, $b = 2$ have interesting properties. This was observed already by Krall and Frink, who suggested in [68] that such a study be undertaken. It seems nevertheless that these polynomials were never investigated thoroughly, although it may be worthwhile to do so. As we have seen in Chapter 14, these polynomials occur in the inversion of the Laplace transform. Do they occur also in some different context? Do they have other practical, or theoretical applications?

2. The BP satisfy a very specific condition, recently called Bessel orthogonality (see [63]). Are there other interesting sets of functions (perhaps even of polynomials) that have this property?

3. In Chapter 9 we studied expansion in series of BP. While the work of Boas and Buck (see [13]) is remarkable, our knowledge of these expansions in series of BP cannot be compared, as completeness, with the theory of expansions in Fourier Series or in series of classical, orthonormal polynomials. Many open problems remain, concerning, e.g., the speed of convergence, behavior of the series on the boundary of its domain of convergence, summability methods (see, however [10] and [96]), Lebesgue constants and other similar questions. The generalizations to the case $a \neq 2$ have, apparently not even been touched.

4. Are there reasonable generalizations of the BP to polynomials in several variables?

5. In Chapter 10, certain regions were determined, where all the zeros of $y_n(z;a)$ (or of $\theta_n(z;a)$ are located (see, e.g. the work of Parodi, Dočev, Saff and Varga, etc). In Section 12, on the other hand, following Olver, a curve is indicated, close to which all these zeros lie, the approximation becoming better with increasing n. There is a gap to be filled between these two types of results. So, e.g., one may determine a strip, enclosing Olver's curve, such that all the zeros of $y_n(z;a)$ belong to that strip.

6. Wimp [112] proved that the zeros $\alpha_j^{(n)}(a)$ $(j = 1,2,\dots,n)$ of $y_n(z;a)$ have strictly negative real parts. One may conjecture that the property holds for a larger set A_1 of values of a. (Conjecture: $A_1 = \mathbb{R}$). Also the set A_2 of values of a, such that $y_n(z;a) = 0$ has a real zero, if n is odd, may deserve study.

7. Determine good upper and lower bounds for the real zero of $y_n(z;a)$, n odd, where $a \in A_2$ (see Problem 6).

8. Find explicit formulae for $\sigma_r^{(n)}$ and $\sigma_{-r}^{(n)}$, the sums of the r-th powers of the zeros of $y_n(z)$, and of $\theta_n(z)$, respectively; if possible, generalize to the case of arbitrary $a \neq 2$.

9. Prove that $y_n(z)$ (hence, also $\theta_n(z)$) is irreducible over the rational field for all n; failing that, prove the irreducibility for a large class of values of n.

10. If possible extend the results of 9 to $y_n(z;a)$, for $a \neq 2$.

11. In Chapter 12 the proof that the Galois group of $y_n(z)$ is the symmetric group on n symbols required almost a case by case consideration for each $n < 13$. The problem is to find a unified approach that works for all these small values of n.

12. Define the function $f_n(z;a)$ by (see (13.9)) $f_n(z;a) = (2nz/e)^{-n}\ 2^{3/2-a}e^{-1/z}y_n(z;a)$; develop $f_n(z;a)$ into a complete asymptotic series.

BIBLIOGRAPHY OF BOOKS AND PAPERS RELATED TO BESSEL POLYNOMIALS

1. W.H. Abdi - *A basic analog of the Bessel Polynomials* - Math. Nachr., vol. 30 (1965), pp. 209-219; MR 32 #7795.

2. R.P. Agarwal - *On Bessel Polynomials* - Canad. J. Math., vol. 6 (1954), pp. 410-415; MR 15-955.

3. W.A. Al-Salam - *The Bessel Polynomials* - Duke Math. J., vol. 24 (1957), pp. 529-545; MR 19-849.

4. W.A. Al-Salam - *On the Bessel Polynomials* - Boll. Un. Mat. Ital. (3), vol. 2 (1957), pp. 227-229, MR 19-542.

5. W.A. Al-Salam - *Some functions related to the Bessel Polynomials* - Duke Math.J., vol. 26 (1959), pp. 519-539; MR 22 #120.

6. W.A. Al-Salam - *Remarks on some operational formulas* - Rend. Sem. Mat. Univ. Padova, vol. 35 (1965), pp. 128-131; MR 31 #4935.

7. W.A. Al-Salam and L. Carlitz - *Bernoulli Numbers and Bessel Polynomials* - Duke Math. J., vol. 26 (1959), pp. 437-445; MR 21 #4256.

8. W.A. Al-Salam and L. Carlitz - *Bessel Polynomials and Bernoulli Numbers* - Arch. Math., vol. 9 (1959), pp. 412-415; MR 21 #3597.

9. W.A. Al-Salam and T.S. Chihara - *Another characterization of classical, orthogonal polynomials* - SIAM J. Math. Anal., vol. 3 (1972), pp. 65-70; MR 47 #5320.

10. I. Baičev - *Convergence and summability of series of generalized Bessel Polynomials* (Bulgarian, Russian and English Summaries) - B"lgar. Akad. Nauk. Otdel. Mat. Fiz. Nauk. Izv. Mat. Inst., vol. 10 (1969), pp. 17-26; MR 44 #5703.

11. D.P. Banerjee - *On Bessel Polynomials* - Proc. Nat. Acad. Sci. India, Sect. A, vol. 29 (1960), pp. 83-86; MR 26 #1505.

12. C.W. Barnes - *Remarks on the Bessel Polynomials* - Amer. Math. Monthly, vol. 80 (1973), pp. 1034-1041; MR 49 #660.

13. R.P. Boas, Jr. and R.C. Buck - *Polynomial expansions of Analytic Functions* - Springer-Verlag, Berlin, 1958; MR 20 #984.

14. S. Bochner - *Über Sturm - Liouvillsche Polynomsysteme* - Math. Z., vol. 29 (1929), pp. 730-736.

15. F. Brafman - *A set of generating functions for Bessel Polynomials* - Proc. Amer. Math. Soc., vol. 4 (1953), pp. 275-277; MR 14-872.

16. J.W. Brown - *On Burchnall's generating relation for Bessel Polynomials* - Amer. Math. Monthly, vol. 74 (1967), pp. 182-183; MR 36 #4034.

17. J. Burchnall - *The Bessel Polynomials* - Canad. J. Math., vol. 3 (1951), pp. 62-68; MR 12-499.

18. J. Burchnall and T.W. Chaundy - *Commutative ordinary differential operators II - The identity* $P^n = Q^m$ - Proc. Royal Soc. Ser A, vol. 134 (1931), pp. 471-485.

19. L. Carlitz - *A note on the Bessel Polynomials* - Duke Math. J., vol. 24 (1957), pp. 151-162; MR 19-27.

20. B.C. Carlson - *Special functions of applied mathematics* - Academic Press, New York - London, 1977.

21. S.K. Chatterjea - *On the Bessel Polynomials* - Rend. Sem. Mat. Univ. Padova, vol. 32 (1962), pp. 295-303; MR 26 #373.

22. S.K. Chatterjea - *A Note on Bessel Polynomials* - Boll. Un. Mat. Ital. (3), vol. 17 (1962), pp. 270-272; MR 26 #1506.

23. S.K. Chatterjea - *Operational formulae for certain classical polynomials I* - Quart. J. Math. Oxford Ser. (2), vol. 14 (1963), pp. 241-246; MR 27 #2662.

24. S.K. Chatterjea - *Operational formulae for certain classical polynomials II* - Rend. Sem. Mat. Univ. Padova, vol. 33 (1963), pp. 163-169; MR 27 #3845.

25. S.K. Chatterjea - *Operational formulae for certain classical polynomials III* - Rend. Sem. Mat. Univ. Padova, vol. 33 (1963), pp. 271-277; MR 27 #3846.

26. S.K. Chatterjea - *An integral representation for the product of two generalized Bessel Polynomials* - Bull. Un. Mat. Ital., vol. 18 (1963), pp. 377-381; MR 28 #5200.

27. S.K. Chatterjea - *A generalization of the Bessel Polynomials* - Mathematica (Cluj), vol. 6 (29) 1 (1964), pp. 19-29.

28. S.K. Chatterjea - *On a paper by Banerjee* - Bull. Un. Mat. Ital., vol. 19 (1964), pp. 140-145; MR 29 #3695.

29. S.K. Chatterjea - *A new class of polynomials* - Ann. Mat. Pura Appl., vol. (4) 65 (1964), pp. 35-48; MR 29 #6073.

30. S.K. Chatterjea - *Some generating functions* - Duke Math. J., vol. 32 (1965), pp. 563-564; MR 31 #5989.

31. S.K. Chatterjea - *Some generating functions of Bessel Polynomials* - Math.Japon., vol. 10 (1965), pp. 27-29; MR 32 #7797.

32. S.K. Chatterjea - *An integral involving Turán's expression for Bessel Polynomials* - Amer. Math. Monthly, vol. 72 (1965), pp. 743-745; MR 32 #2630.

33. S.K. Chatterjea - *Operational derivation of some results for Bessel Polynomials* Mat. Vesnik, vol. 3 (18), 1966, pp. 176-186; MR 34 #7833.

34. S.K. Chatterjea - *On Turán's expression for Bessel Polynomials* - Bangabasi Morning College Mag., vol. 2 (1966), pp. 18-19; MR 35 #6876.

35. S.K. Chatterjea - *Sur les polynômes de Bessel, du point de vue de l'algèbre de Lie* - C. R. Acad. Sci. Paris, Serie A, vol. 271 (1970), pp. 357-360; MR 42 #3329.

36. S.K. Chatterjea - *Operational derivation of some generating functions for the Bessel Polynomials* - Math. Balkanica, vol. 1 (1971), pp. 292-297; MR 44 #7008.

37. S.K. Chatterjea - *Some properties of simple Bessel Polynomials from viewpoint of Lie algebra* - C. R. Acad. Bulgare Sci., vol. 28 (1975), pp. 1455-1458; (not reviewed in MR up to 1977).

38. C.K. Chatterjee - *On Bessel Polynomials I.* - Bull. Calcutta Math. Soc., vol. 49 (1957), pp. 67-70; MR 20 #3308.

39. M.P. Chen and C.C. Feng - *Group theoretic origins of certain generating functions for generalized Bessel Polynomials* - Tamkang J. Math., vol. 6 (1975), pp. 87-93; MR 51 #8495.

40. J.A. Cima - *Note on a theorem of Grosswald* - Trans. Amer. Math. Soc., vol. 99 (1961), pp. 60-61; MR 22 #11156.

41. M.K. Das - *Operational representations for the Bessel Polynomials* - Bull. Soc. Math. Phys. Macédoine, vol. 17 (1966), pp. 27-32 (1968); MR 39 #501.

42. M.K. Das - *A generating function for the general Bessel Polynomial* - Amer. Math Monthly, vol. 74 (1967), pp. 182-183; MR 34 #7843.

43. M.K. Das - *Sur les polynômes de Bessel, du point de vue de l'algèbre de Lie* - C. R. Acad. Sci. Paris, Ser. A., vol. 271 (1970), pp. 361-364; MR 42 #3330.

44. M.K. Das - *Sur les polynômes de Bessel* - C. R. Acad. Sci. Paris, Ser. A, vol. 271 (1970), pp. 408-411; MR 42 #6300.

45. M.K. Das - *Operational formulas connected with some classical orthogonal polynomials* - Bull. Math. Soc. Sci. Math. R.S. Roumaine (N.S.) vol. 14 (62) (1970), pp. 283-291; MR 49 #10936.

46. G.K. Dhawan and D.D. Paliwal - *Generating functions of Gegenbauer, Bessel and Laguerre Polynomials* - Math. Education, vol. 10 (1976), pp. A9-A15; MR 53 #5963.

47. D. Dickinson - *On Lommel and Bessel Polynomials* - Proc. Amer. Math. Soc., vol. 5 (1954), pp. 946-956; MR 19-263.

48. K. Dočev - *On the generalized Bessel Polynomials* - Bulgar. Akad. Nauk. Izv. Mat. Inst., vol. 6 (1962), pp. 89-94; MR 26 #2645.

49. M. Dutta, S.K. Chatterjea, M.L. Moore - *On a class of generalized Hermite polynomials* - Bull. Inst. Math. Acad. Sinica, vol. 3 (1975), No. 2, pp. 377-381; MR 52 #11149.

50. A. Erdélyi, W. Magnus, F. Oberhettinger, F.G. Tricomi - *Higher transcendental functions* (3 volumes) (Based in part on notes left by H. Bateman) - McGraw-Hill Book Co., New York - Toronto - London, 1953; MR 15-419 and MR 16-586.

51. M.T. Eweida - *On Bessel Polynomials* - Math. Z., vol. 74 (1960), pp. 319-324; MR 22 #8153.

52. M.T. Eweida - *Infinite integrals involving Bessel Polynomials* - Univ. Nac. Tucuman Rev. Ser. A, vol. 13 (1960), pp. 132-135; MR 24(A) #262.

53. E. Grosswald - *On some algebraic properties of the Bessel Polynomials* - Trans. Amer. Math. Soc., vol. 71 (1951), pp. 197-210; MR 14-747.

54. E. Grosswald - *Addendum to "On some algebraic properties of the Bessel Polynomials"* - Trans. Amer. Math. Soc., vol. 144 (1969), pp. 569-570; MR 40 #4246.

55. E. Grosswald - *The student t-distribution for odd degrees of freedom is infinitely divisible* - Ann. Probability, vol. 4 (1976), pp. 680-683; MR 53 #14591.

56. E. Grosswald - *The student t-distribution of any degree of freedom is infinitely divisible* - Z. Wahrescheinlichkeitstheorie und verw. Gebiete, vol. 36 (1976), pp. 103-109; MR 54 #14037.

57. E.A. Guillemin - *Synthesis of passive Networks* - John Wiley & Sons, Inc., New York; Chapman & Hill, Ltd., London, 1958; MR 25 #932.

58. W. Hahn - *Ueber die Jacobischen Polynome und zwei verwandte Polynomklassen* - Math. Z., vol. 39 (1935), pp. 634-638.

59. A.M. Hamza - *Integrals involving Bessel Polynomials* - Proc. Math. Phys. Soc. A.R.E., vol. 35 (1971), pp. 9-15; MR 48 #8888.

60. D. Hazony - *Elements of network synthesis* - Reinhold Publ. Corp., New York; Chapman & Hall, Ltd. London, 1963.

61. M.E.H. Ismail - *Dual and triple sequence equations involving orthogonal polynomials* - Nederl. Akad. Wetensch. Proc. Ser. A, vol. 78 - Indag. Math., vol. 37 (1975), pp. 164-169; MR 52 #3901.

62. M.E.H. Ismail and D.N. Kelker - *The Bessel Polynomials and the Student t-distribution* - SIAM J. Math. Anal., vol. 7 (1976), pp. 82-91; MR 52 #12164.

63. J.W. Jayne - *Polynomials orthogonal on a contour - the Bessel alternative* - (to appear).

64. Johnson, Johnson, Boudra, Stokes - *Filters using Bessel type polynomials* - I.E.E.E. Trans. Circuits and Systems, vol. CAS-23, No. 2, February 1976, pp. 96-99.

65. D.N. Kelker - *Infinite divisibility and various mixtures of the normal distribution* - Ann. Math. Statist., vol. 42 (1971), pp. 802-808; MR 44 #3415.

66. A.M. Krall - *Orthogonal polynomials through moments generating functionals* - SIAM J. Math. Anal., vol. 9 (1978), pp. 600-603.

67. H.L. Krall - *On derivatives of orthogonal polynomials II* - Bull. Amer. Math. Soc., vol. 47 (1941), pp. 261-264; MR 2-282.

68. H.L. Krall and O. Frink - *A new class of orthogonal polynomials: the Bessel Polynomials* - Trans. Amer. Math. Soc., vol. 65 (1949), pp. 100-115; MR 10-453.

69. V.I. Krylov and N.S. Skoblia - *Handbook on the numerical inversion of the Laplace transform (Russian)* - Izdat. "Nauka i Tehnika", Minsk, 1968, 295 pages; MR 38 #1814.

70. V.N. Kublanowskaya and T.N. Smirnova - *Zeros of Hankel functions and other functions associated with them (Russian)* - Trudy Mat. Inst. Steklov, vol. 53 (1959), pp. 186-191; MR 22 #1105.

71. Y.L. Luke - *Special Functions and their Approximations* (3 volumes), Academic Press, Inc., New York - London, 1969; MR 39 #3039 and MR 40 #2909.

72. Y.L. Luke - *Mathematical Functions and their Approximations* - Academic Press, Inc., New York - San Francisco - London, 1975.

73. E.B. McBride - *Obtaining generating functions - Springer Tracts in Natural Philosophy*, vol. 21, Springer-Verlag, New York - Heidelberg, 1971; MR 43 #5077.

74. P.J. McCarthy - *Approximate location of the zeros of generalized Bessel Polynomials* - Quart. J. Math., Oxford, Ser. (2), vol. 12 (1961), pp. 265-267; MR 25 #243.

75. P.J. McCarthy - *Characterizations of classical polynomials* - Portugal. Math., vol. 20 (1961), pp. 47-52; MR 23 #A3866.

76. H.B. Mittal - *Some generating functions* - Rev. da Faculdade de Ciencias de Lisboa, Ser. 2, vol. 13 (1970), pp. 43-45; MR 46 #7600.

77. H.B. Mittal - *Polynomials defined by generating functions* - Trans. Amer. Math. Soc., vol. 168 (1972), pp. 75-84; MR 45 #3811.

78. H.B. Mittal - *Some generating functions for polynomials* - Czechoslovak Math.J., vol. 24 (99) (1974), pp. 341-348; MR 50 #660.

79. H.B. Mittal - *Unusual generating relations for polynomial sets* - J. Reine Angew. Math., vol. 271 (1974), pp. 122-137; MR 51 #949.

80. R.D. Morton and A.M. Krall - *Distributional weight functions for orthogonal polynomials* - SIAM J. Math. Anal., vol. 9 (1978), pp. 604-626.

81. M. Nasif - *Note on the Bessel Polynomials* - Trans. Amer. Math. Soc., vol. 77 (1954), pp. 408-412; MR 16 #818.

82. N. Obreshkov - *About certain orthogonal polynomials in the complex plane* - B"lgar. Akad. Nauk Izv. Math. Inst., vol. 2 (2) (1956), pp. 45-68.

83. C.D. Olds - *The simple continued fraction expansion of e* - Amer. Math. Monthly, vol. 77 (1970), pp. 968-975.

84. F.W.J. Olver - *The asymptotic expansions of Bessel Functions of large order* - Philos. Trans. Roy. Soc. London, Sec. A, vol. 247 (1954/5), pp. 328-368; MR 16-696.

85. M. Parodi - *Sur les Polynômes de Bessel* - C. R. Acad. Sci. Paris, vol. 274A (1972), pp. A1153-A1155; MR 46 #416.

86. A. Pham - Ngoc Dinh - *Un nouveau type de dévelopment des polynômes généralisés de Bessel* - C. R. Acad. Sci Paris, Ser. A, vol. 272 (1971), pp. A1393-A1396; MR 44 #2959.

87. R. Piessens - *Gaussian quadrature formulas for the numerical integration of Bromwich's integral and the inversion of the Laplace transform* - J.Engrg. Math., vol. 5 (1971), pp. 1-9; MR 42 #8664.

88. F.M. Ragab - *Series of products of Bessel Polynomials* - Canad. J. Math., vol. 11 11 (1959), pp. 156-160; MR 21 #733.

89. E.D. Rainville - *Generating functions for Bessel and related polynomials* - Canad. J. Math., vol. 5 (1953), pp. 104-106; MR 14-872.

90. E.D. Rainville - *Special Functions* - The Macmillan Co., New York, 1960; MR 21 #6447.

91. A.K. Rajagopal - *On Bessel Polynomials* - Bull. Un. Mat. Ital., vol. (3) 13 (1958), pp. 418-422; MR 21 #1408.

92. V. Romanovsky - *Sur quelques classes nouvelles des polynômes orthogonaux* - C.R. Acad. Sci. Paris, vol. 188 (1929), pp. 1023-1025.

93. H. van Rossum - (a) *A theory of orthogonal polynomials, based on the Padé Table (Thesis)*, University of Utrecht, 76 pages - van Gorcum, Assen, 1953;
(b) *Systems of orthogonal and quasi orthogonal polynomials connected with the Padé Table, I, II, III* - Nederl. Akad. Wetensch. Proc. Ser. A, vol. 58 - Indag. Math., vol. 17 (1955), pp. 517-525; 526-534; 675-682; MR 19-412.

94. H. van Rossum - *Totally positive polynomials* - Nederl. Akad. Wetensch., Proc. Ser. A, vol. 68 - Indag. Math., vol. 27 (1965), pp. 305-315; MR 35 #6880.

95. H. van Rossum - *A note on the location of the zeros of the generalized Bessel Polynomials and totally positive polynomials* - Niew Arch. Wisk., vol. (3) 17 (1969), pp. 142-149; MR 41 #534.

96. P. Rusev - *Convergence of series of Jacobi and Bessel Polynomials on the boundaries of their regions of convergence (Bulgarian; Russian and English summaries)* - B"lgar. Akad. Nauk. Otdel. Mat. Fiz. Nauk. Izv. Mat. Inst., vol. 10 (1969), pp. 17-26; MR 34 #8074.

97. H. Rutishauser - *Bemerkungen zu einer Arbeit von Al-Salam und Carlitz* - Arch. Math., vol. 10 (1959), pp. 292-293; MR 21 #4262.

98. E.B. Saff and R.S. Varga - *Zerofree parabolic regions for sequences of polynomials* - SIAM J. Math. Anal., vol. 7 (1976), pp. 344-357; MR 54 #3060.

99. H.E. Salzer - *Orthogonal polynomials arising in the numerical evaluation of inverse Laplace transforms* - Mathematical Tables and other Aids to Computation (MTAC), vol. 9 (1955), pp. 164-177; MR 17-1203.

100. H.E. Salzer - *Additional formulas and tables for orthogonal polynomials originating from inversion integrals* - J. Math. and Phys., vol. 40 (1961), pp. 72-86; MR 23B #B2612.

101. R.R. Shepard - *Active filters: Part 12 - Shortcut to network design* - Electronics, August 1969, pp. 82-91.

102. N. Skoblja - *Tables for the numerical inversion of the Laplace transforms*
$$f(x) = \frac{1}{2\pi i}\int_{c-i\infty}^{c+i\infty} e^{xp}\, F(p)\,dp$$
- Izdat. "Nauka i Tehnika", Minsk, 1964, 44 pages; MR 29 #725.

103. N. Skoblja - *The distribution of zeros of polynomials connected with the numerical integration of the Laplace transform (Russian)* - Dokl. Akad. Nauk BSSR 9 (1965), pp. 288-291; MR 32 #4841.

104. S.L. Soni - *A note on Bessel Polynomials* - Proc. Indian Acad. Sci. Sect. A, vol. 71 (1970), pp. 93-99; MR 41 #7166.

105. H.M. Srivastava - *On Bessel, Jacobi and Laguerre Polynomials* - Rend.Sem. Mat. Univ. Padova, vol. 35 (1965), pp. 424-432; MR 33 #4346.

106. L. Storch - (a) *Synthesis of constant-time-delay ladder network using Bessel Polynomials* - Proc. IRE, vol. 42, pp. 1666-1676, Nov. 1954;
(b) *An application of modern network synthesis to the design of constant-time-delay networks with low-q elements* 1954 IRE Convention Record, Part 2, Circuit Theory, pp. 105-117; MR 16-1182.

107. W.E. Thomson - *Delay network having maximally flat frequency characteristics* - Proc. Instituto Electr. Engineers, vol. 96 (1949), Part III, p. 487.

108. W.E. Thomson - *Networks with maximally flat delay* - Wireless Engineer, vol. 29 (1952), pp. 256-263.

109. L. Toscano -*Funzioni generatrici di particolari polinomi di Laguerre e di altri da essi dipendenti* - Bull. Un. Mat. Ital., vol.(3) 7 (1952), pp. 160-167; MR 14, p. 269.

110. L. Toscano - *Osservazioni, confronti e complementi su particolari polinomi ipergeometrici* - Le Matematiche, vol. 10 (1955), pp. 121-133; MR 17, p. 733.

111. H.S. Wall - *Polynomials whose zeros have negative real parts* - Amer. Math. Monthly, vol. 52 (1945), pp. 308-325; MR 7-62.

112. J. Wimp - *On the zeros of a confluent hypergeometric function* - Proc. Amer. Math. Soc., vol. 16 (1965), pp. 281-283; MR 30 #4001.

113. A. Wragg and C. Underhill - *Remarks on the zeros of Bessel Polynomials* - Amer. Math. Monthly, vol. 83 (1976), pp. 122-126; MR 52 #1146.

SUPPLEMENTARY BIBLIOGRAPHY OF TITLES OBTAINED AFTER COMPLETION OF THE PRESENT MONOGRAPH.

114. W. Miller, Jr. - *Encyclopedia of mathematics and its applications*, vol. 4, *Symmetry and separation of variables* - Addison-Wesley Publ. Co., Reading, Mass., 1977.

115. L. Weinberg - *Network analysis and synthesis* - McGraw-Hill Book Co., Inc., New York, 1955; Krieger Publishing Co., Huntington, N.Y., 1975.

116. A.H. Marshak, D.E. Johnson and J.R. Johnson - *A Bessel rational filter* - IEEE Trans. Circ. Supt. vol. CAS - 21 (1974), 797-799.

GENERAL LITERATURE, NOT DIRECTLY RELATED TO BESSEL POLYNOMIALS

1. M. Abramovitz and I.E. Segun - *Handbook of Mathematical Functions* - Dover Publications, Inc. New York, 1968.

2. R. Askey - *Dual equations and classical orthogonal polynomials* - J. Math. Anal. Appl., vol. 24 (1968), pp. 677-685.

3. I. Bendixson - *Sur les racines d'une équation fondamentale* - Acta Math., vol. 25 (1902), pp. 359-365.

4. R.P. Boas - *Entire functions* - Academic Press, New York, 1954.

5. R.P. Boas and R.C. Buck - *Polynomials defined by generating relations* - Amer. Math Monthly, vol. 63 (1956), pp. 626-632.

6. W.N. Bailey - *Generalized hypergeometric series* - *Cambridge Tracts in Mathematics and Mathematical Physics*, No. 32 - Cambridge University Press, Cambridge (England), 1935.

7. L. Bondesson - *A general result on infinite divisibility* (to appear).

8. R. Breusch - *Zur Verallgemeinerung des Bertrandschen Postulates* - Math. Z., vol. 34 (1932), pp. 505-524.

9. W. Burnside - *The theory of groups of finite orders, 2nd edition* - Cambridge University Press, Cambridge (England), 1911.

10. M.L. Cartwright - *Integral Functions* - *Cambridge tracts in Mathematics and Mathematical Physics*, No. 44 - Cambridge University Press, Cambridge (England), 1962.

11. H. Davenport - *The higher arithmetic* - Harper Torchbook - Harper & Brothers, New York, 1960.

12. M.G. Dumas - *Sur quelques cas d'irréductibilité des polynômes à coefficients rationels* - J. Math. Pures Appl. (6), vol. 2 (1906), pp. 191-258.

13. G. Doetsch - *Theorie und Anwendung der Laplace Transformation* - Springer Verlag, Berlin (1937).

14. G. Eneström - *Remarques sur un théorème relatif aux racines de l'équation...* - Tohoku Math. J., vol. 18 (1920), pp. 34-36.

15. L. Euler - *Introductio analysin infinitorium, Lausanne* 1 (1748), Chapter 18 - Commentarii Acad. Scientiarum Imperialis Petropolitanae, vol. 9 (1737) - Operae Ser. I, vol. 18.

16. E. Feldheim - *Relations entre les polynômes de Jacobi, Laguerre et Hermite* - Acta Math., vol. 74 (1941), pp. 117-138.

17. W. Feller - *An introduction to probability theory and its applications (2 vol.), 3rd edition* - John Wiley & Sons, New York, 1967.

18. L.R. Ford - *Differential equations, 2nd edition* - McGraw Hill Book Co., Inc., New York, 1955.

19. B.V. Gnedenko and A.N. Kolmogorov - *Limit distributions for sums of independent random variables* - Addison-Wesley Publ. Co., Inc., Cambridge (Mass.), 1954.

20. H.W. Gould - *Combinatorial identies* - Morgantown, W. Va., 1972.

21. H.W. Gould - *Generalization of binomial identies of Carlitz, Grosswald and Riordan* (to appear).

22. H.W. Gould - *Private communication* (letter of 9 August 1977).

23. E. Grosswald - *On a simple property of the derivatives of Legendre Polynomials* - Proc. Amer. Math. Soc., vol. 1 (1950), pp. 553-554.

24. E. Grosswald - *Topics from the theory of numbers* - The Macmillan Co., New York, 1966.

25. E. Grosswald and H.L. Krall - *Evaluation of two determinants* - Amer. Math. Monthly (Advanced Problems; to appear).

26. H. Hamburger - *Beiträge zur Konvergenztheorie der Stieltjesschen Kettenbrüche* - Math. Z., vol. 4 (1919), pp. 186-222.

27. H. Hamburger - *Ueber eine Erweiterung des Stieltjesschen Momentproblems* - Math. Ann., vol. 81 (1920), pp. 235-319; vol. 82 (1920), pp. 120-164 and pp. 168-187.

28. K. Hensel and O. Landsberg - *Theorie der algebraischen Funktionen* - Teubner, Leipzig, 1902.

29. C. Hermite - *Sur la fonction exponentielle* - C. R. Acad. Sci. Paris, vol. 77 (1873), pp. 18-24; 74-79; 226-233; 285-293; also Oeuvres, vol. 3, pp. 150-181.

30. I.N. Herstein - *Topics in algebra* - Blaisdell Publ. Co., Waltham (Mass.), 1964.

31. I.N. Herstein - *Private communication* (letter of 21 November 1951).

32. E. Hille - *Analytic function theory* - Ginn & Co., Boston, 1962.

33. M.A. Hirsch - *Sur les racines d'une équation fondamentale* - Acta Math., vol. 25 (1902), pp. 367-370.

34. A. Hurwitz - *Ueber einen Satz des Herrn Kakeya* - Tohoku Math. J., vol. 4 (1913), pp. 89-93.

35. M.E.H. Ismail - *Bessel functions and the infinite divisibility of the Student t-distribution* - Ann. Probability, vol. 5 (1977), pp. 582-585.

36. M.E.H. Ismail - *Integral representations and complete monotonicity of various quotients of Bessel functions* - Canad. J. Math., vol. 29 (1977), pp. 1198-1207.

37. K.E. Iverson - *The zeros of the partial sums of* e^z - Math Tables and other Aids to Computation (MTAC), vol. 7 (1953), pp. 165-168.

38. D. Jackson - *Fourier Series and orthogonal polynomials* - *Carus Monograph No. 6*, The Math Assoc. of America, Oberlin (Ohio), 1941.

39. C. Jordan - *Sur la limite de transitivité des groupes non alternés* - Bull. Soc. Math. France, vol. 1 (1872/3), pp. 40-71.

40. S. Kakeya - *On the limits of the roots of an algebraic equation* - Tohoku Math. J., vol. 2 (1912), pp. 140-142.

41. E. Laguerre - *Sur la distribution dans le plan des racines d'une équation algébrique, dont le premier membre satisfait à une équation lineaire du second ordre* - C.R. Acad. Sci. Paris, vol. 94 (1882), pp. 412-416; 508-510;

Oeuvres, vol. 1, pp. 161-166.

42. J.H. Lambert - *Mémoire sur quelques propriétés remarquables des quantités transcendentes, circulaires et logarithmiques* - Histoire de l'Acad. royale des sciences et belles-lettres - Berlin, 1761 and 1768.

43. E. Landau - *Handbuch der Lehre von der Verteilung der Primzahlen* - Teubner, Leipzig, 1909.

44. P. Lévy - *Théorie de l'addition des variables aléatoires* - Gauthier-Villars, Paris, 1937.

45. M. Marcus and H. Minc - *A survey of matrix theory and matrix inequalities* - Prindle, Weber and Schmidt, Boston, 1964.

46. F.W.J. Olver - *The asymptotic solution of linear differential equations of the second order for large values of a parameter* - Philos. Trans. Roy. Soc. London, Ser. A, vol. 247 (1954/5), pp. 307-327.

47. O. Perron - *Die Lehre von den Kettenbrüchen* - Teubner, Leipzig, 1929.

48. N. du Plessis - *A note about the derivatives of Legendre's Polynomials* - Proc. Amer. Math. Soc., vol. 2 (1951), p. 950.

49. G. Pólya and G. Szegö - *Aufgaben und Lehrsätze aus der Analyse* - Julius Springer, Berlin, 1925.

50. E. Rainville - *On a simple property of the derivatives of Legendre's Polynomials* - Unpublished manuscript (letter of 21 October 1950).

51. I. Schur - *Einige Sätze über Primzahlen* - Sitzungsberichte der Preussischen Akad. der Wissenschaften (1929), pp. 125-136; Gesammelte Abhandlungen, vol. III, No. 64, pp.140-151.

52. I. Schur - *Gleichungen ohne Affekt* - Sitzungsberichte der Preussischen Akad. der Wissenschaften (1930), pp. 443-449; Gesammelte Abhandlungen, vol. III, No. 67, pp. 191-197.

53. C.L. Siegel - *Transcendental Numbers* - *Annals of Mathem. Studies No. 16* - Princeton University Press, Princeton, 1949.

54. T.J. Stieltjes - *Recherches sur les fractions continues* - Annales de la Fac. des Sci. Toulouse, vol. 8 (1894), 122 pages and vol. 9 (1895), 47 pages; Oeuvres Completes, vol. 2, pp. 402-566.

55. J.A. Stratton - *Electromagnetic Theory* - McGraw Hill, New York, 1947.

56. G. Szegö - *Orthogonal polynomials* - *Colloquium Publication*, vol. 23 - Amer. Math. Soc., New York, 1939.

57. P.L. Tchebycheff - *Sur l'Interpolation* - Zapiski Akad. Nauk., vol. 4, Supplement 5 (1864); Oeuvres, vol. 11, pp. 539-590.

58. O. Thorin - *On the infinite divisibility of the log normal distribution* - Scand. Actuar. J. (1977), pp. 121-148.

59. B.L. van der Waerden - *Modern Algebra* (Revised English edition) - Frederick Ungar Publ. Co., New York, 1949.

60. J.H. Wahab - *New cases of irreducibility for Legendre Polynomials*-Duke Math.J., vol. 19 (1952), pp. 165-176.

61. G.N. Watson - *A treatise on the theory of Bessel Functions, 2nd edition* - Cambridge University Press, Cambridge (England), 1946.

62. L. Weisner - *Group theoretic origin of certain generating functions* - Pacific J. Math., vol. 5 (1955), pp. 1033-1039.

63. L. Weisner - *Generating functions for Hermite functions* - Canad. J. Math., vol. 11 (1959), pp. 141-147.

64. L. Weisner - *Generating functions for Bessel functions* - Canad. J. Math., vol. 11 (1959), pp. 148-155.

65. F.J.W. Whipple - *Some transformations of generalized hypergeometric series* - Proc. London Math. Soc. (2), vol. 26 (1927), pp. 257-272.

66. E.J. Whittaker - *Sur les séries de base de polynômes quelconques* - Gauthier-Villars, Paris, 1949.

67. E.J. Whittaker and G.N. Watson - *Modern Analysis, 4th edition* - Cambridge University Press, Cambridge (England), 1927.

68. D.V. Widder - *The Laplace Transform* - Princeton University Press, Princeton, 1941.

69. H. Wilf - *Mathematics for Physical Sciences* - J. Wiley & Sons, Inc., New York and London, 1962.

SUBJECT INDEX

The numbers following each entry refer to page numbers
I stands for any page in the Introduction.

NAME INDEX

The numbers following each name stand for the page numbers; the letters F and I stand for any page of the Foreword, or of the Introduction, respectively.

PARTIAL LIST OF SYMBOLS

$\mathbb{C}$	- field of complex numbers
$\mathbb{Z}$	- the rational integers
$\mathbb{Z}^+$	- the positive integers
$J_\nu(z), I_\nu(z), K_\nu(z)$	- standard notations for Bessel and MacDonald functions
$H_\nu^{(1)}(z), H_\nu^{(2)}(z)$	- standard notations for Hankel functions
$y_n(z), y_n(z;a,b)$	- Bessel Polynomial, simple and generalized, standardization of Krall and Frink.
$\theta_n(z), \theta_n(z;a,b)$	- Bessel Polynomial, simple and generalized, standardization of Burchnall.
$P_n^{(a,b)}(z)$	- Jacobi Polynomial
$L_n(x)$	- Legendre Polynomial
$L_n^{(a)}(z)$	- Laguerre Polynomial
$H_n(z)$	- Hermite Polynomial
$R_{n,\nu}(z)$	- Lommel Polynomial
$P_n^q(z)$	- associate Legendre function
$W_{k,m}(z)$	- Whittaker function
${}_pF_q(a_1,\dots,a_p;b_1,\dots,b_q;z)$	- generalized hypergeometric function
$\Gamma(z)$	- gamma function
$B(u,v)$	- beta function
$[x]$	- greatest integer function
$\mathrm{erfc}(z)$	- complementary error function
$w(x), \rho(x;a,b)$	- notations for weight functions
$T(s)$	- transfer function
$Z(s)$	- driving point impedance
$A_n(s)$	- distortion factor
$A_n = \lvert A_n(i\omega t_o)\rvert$	- size of the distortion factor for $s = i\omega t_o$
Δ	- Laplacian operator
δ	- differential operator $x\frac{d}{dx}$
$\underline{R}$	- right shift operator
$\underline{L}$	- left shift operator
$L(x, \frac{d}{dx}, \alpha)$	- linear, ordinary differential operator that depends also linearly on a parameter α
Δ_a	- difference operator with respect to the parameter a.
$L(f)(t)$	- the Laplace transform of $f(x)$ at t
$L^{-1}(g)(x)$	- the inverse Laplace transform of $g(t)$ at x
$M_k(f,\rho)$	- k-th moment of the function f with weight function ρ

$\binom{m}{n}$	- binomial coefficient
$[a_o,a_1,\ldots,a_n,\ldots]$	- continued fraction
B_n	- n-th Bernoulli number
$A_i^{(n)}$	- Christoffel constants
$S_n^{(m)}$	- Stirling numbers of the second kind
δ_{ij}	- takes only the values 0, or 1; when used specifically as a Kronecker delta ($\delta_{ij} = 1$ if and only if $i = j$), this is stated explicitly.
$\alpha_k^{(n)}, \alpha_k^{(n)}(a,b)$	- k-th zero of $y_n(z)$, or of $y_n(z;a,b)$, respectively
$\beta_k^{(n)}, \beta_k^{(n)}(a,b)$	- k-th zero of $\theta_n(z)$, or of $\theta_n(z;a,b)$, respectively
$\sigma_r^{(n)}$	- sum of r-th powers of the zeros of $y_n(z)$
(a,b)	- greatest common divisor of a and b
$[a,b]$	- spot of coordinates a,b
$a \mid b$	- a divides b
$e_m(p)$	- the exact power of p that divides m
$a \equiv b$	- a is congruent b with respect to some modulus
$f \sim g$	- asymptotic, or formal equality
$f = O(g)$	- means $\lvert f(z)/g(z) \rvert < C$, some constant
P_n	- permutation on n symbols
$(a,b,c)(d,e)$	- cycles of a permutation; also used for a scheme of factorization.
N.p.	- Newton polygon
A^*	- complex conjugate of the matrix A
λ_j,μ_j,ν_j	- notations for characteristic (= eigen) values of a matrix
$R(f,g)$	- resultant of the polynomials f and g
D, D_n	- denotes usually a discriminant, D also a differentiation operator
$\sigma(n), \sigma_p(n)$	- sum of the p-adic digits of n
ζ_z	- center of mass with respect to the pole z
$a_m, a_m^{(n)}$	- m-th coefficient of the n-th polynomial, often of $y_n(z)$, or $\theta_n(z)$
$a_m, (a)_m$, or $a_{(m)}$	- abbreviation for $a(a+1)\ldots(a+m-1)$
$a^{(m)}$	- abbreviation for $a(a-1)\ldots(a-m+1)$

Vol. 521: G. Cherlin, Model Theoretic Algebra – Selected Topics. IV, 234 pages. 1976.

Vol. 522: C. O. Bloom and N. D. Kazarinoff, Short Wave Radiation Problems in Inhomogeneous Media: Asymptotic Solutions. V. 104 pages. 1976.

Vol. 523: S. A. Albeverio and R. J. Høegh-Krohn, Mathematical Theory of Feynman Path Integrals. IV, 139 pages. 1976.

Vol. 524: Séminaire Pierre Lelong (Analyse) Année 1974/75. Edité par P. Lelong. V, 222 pages. 1976.

Vol. 525: Structural Stability, the Theory of Catastrophes, and Applications in the Sciences. Proceedings 1975. Edited by P. Hilton. VI, 408 pages. 1976.

Vol. 526: Probability in Banach Spaces. Proceedings 1975. Edited by A. Beck. VI, 290 pages. 1976.

Vol. 527: M. Denker, Ch. Grillenberger, and K. Sigmund, Ergodic Theory on Compact Spaces. IV, 360 pages. 1976.

Vol. 528: J. E. Humphreys, Ordinary and Modular Representations of Chevalley Groups. III, 127 pages. 1976.

Vol. 529: J. Grandell, Doubly Stochastic Poisson Processes. X, 234 pages. 1976.

Vol. 530: S. S. Gelbart, Weil's Representation and the Spectrum of the Metaplectic Group. VII, 140 pages. 1976.

Vol. 531: Y.-C. Wong, The Topology of Uniform Convergence on Order-Bounded Sets. VI, 163 pages. 1976.

Vol. 532: Théorie Ergodique. Proceedings 1973/1974. Edité par J.-P. Conze and M. S. Keane. VIII, 227 pages. 1976.

Vol. 533: F. R. Cohen, T. J. Lada, and J. P. May, The Homology of Iterated Loop Spaces. IX, 490 pages. 1976.

Vol. 534: C. Preston, Random Fields. V, 200 pages. 1976.

Vol. 535: Singularités d'Applications Differentiables. Plans-sur-Bex. 1975. Edité par O. Burlet et F. Ronga. V, 253 pages. 1976.

Vol. 536: W. M. Schmidt, Equations over Finite Fields. An Elementary Approach. IX, 267 pages. 1976.

Vol. 537: Set Theory and Hierarchy Theory. Bierutowice, Poland 1975. A Memorial Tribute to Andrzej Mostowski. Edited by W. Marek, M. Srebrny and A. Zarach. XIII, 345 pages. 1976.

Vol. 538: G. Fischer, Complex Analytic Geometry. VII, 201 pages. 1976.

Vol. 539: A. Badrikian, J. F. C. Kingman et J. Kuelbs, Ecole d'Eté de Probabilités de Saint Flour V-1975. Edité par P.-L. Hennequin. IX, 314 pages. 1976.

Vol. 540: Categorical Topology, Proceedings 1975. Edited by E. Binz and H. Herrlich. XV, 719 pages. 1976.

Vol. 541: Measure Theory, Oberwolfach 1975. Proceedings. Edited by A. Bellow and D. Kölzow. XIV, 430 pages. 1976.

Vol. 542: D. A. Edwards and H. M. Hastings, Čech and Steenrod Homotopy Theories with Applications to Geometric Topology. VII, 296 pages. 1976.

Vol. 543: Nonlinear Operators and the Calculus of Variations, Bruxelles 1975. Edited by J. P. Gossez, E. J. Lami Dozo, J. Mawhin, and L. Waelbroeck, VII, 237 pages. 1976.

Vol. 544: Robert P. Langlands, On the Functional Equations Satisfied by Eisenstein Series. VII, 337 pages. 1976.

Vol. 545: Noncommutative Ring Theory. Kent State 1975. Edited by J. H. Cozzens and F. L. Sandomierski. V, 212 pages. 1976.

Vol. 546: K. Mahler, Lectures on Transcendental Numbers. Edited and Completed by B. Diviš and W. J. Le Veque. XXI, 254 pages. 1976.

Vol. 547: A. Mukherjea and N. A. Tserpes, Measures on Topological Semigroups: Convolution Products and Random Walks. V, 197 pages. 1976.

Vol. 548: D. A. Hejhal, The Selberg Trace Formula for PSL (2, ℝ). Volume I. VI, 516 pages. 1976.

Vol. 549: Brauer Groups, Evanston 1975. Proceedings. Edited by D. Zelinsky. V, 187 pages. 1976.

Vol. 550: Proceedings of the Third Japan – USSR Symposium on Probability Theory. Edited by G. Maruyama and J. V. Prokhorov. VI, 722 pages. 1976.

Vol. 551: Algebraic K-Theory, Evanston 1976. Proceedings. Edited by M. R. Stein. XI, 409 pages. 1976.

Vol. 552: C. G. Gibson, K. Wirthmüller, A. A. du Plessis and E. J. N. Looijenga. Topological Stability of Smooth Mappings. V, 155 pages. 1976.

Vol. 553: M. Petrich, Categories of Algebraic Systems. Vector and Projective Spaces, Semigroups, Rings and Lattices. VIII, 217 pages. 1976.

Vol. 554: J. D. H. Smith, Mal'cev Varieties. VIII, 158 pages. 1976.

Vol. 555: M. Ishida, The Genus Fields of Algebraic Number Fields. VII, 116 pages. 1976.

Vol. 556: Approximation Theory. Bonn 1976. Proceedings. Edited by R. Schaback and K. Scherer. VII, 466 pages. 1976.

Vol. 557: W. Iberkleid and T. Petrie, Smooth S^1 Manifolds. III, 163 pages. 1976.

Vol. 558: B. Weisfeiler, On Construction and Identification of Graphs. XIV, 237 pages. 1976.

Vol. 559: J.-P. Caubet, Le Mouvement Brownien Relativiste. IX, 212 pages. 1976.

Vol. 560: Combinatorial Mathematics, IV, Proceedings 1975. Edited by L. R. A. Casse and W. D. Wallis. VII, 249 pages. 1976.

Vol. 561: Function Theoretic Methods for Partial Differential Equations. Darmstadt 1976. Proceedings. Edited by V. E. Meister, N. Weck and W. L. Wendland. XVIII, 520 pages. 1976.

Vol. 562: R. W. Goodman, Nilpotent Lie Groups: Structure and Applications to Analysis. X, 210 pages. 1976.

Vol. 563: Séminaire de Théorie du Potentiel. Paris, No. 2. Proceedings 1975–1976. Edited by F. Hirsch and G. Mokobodzki. VI, 292 pages. 1976.

Vol. 564: Ordinary and Partial Differential Equations, Dundee 1976. Proceedings. Edited by W. N. Everitt and B. D. Sleeman. XVIII, 551 pages. 1976.

Vol. 565: Turbulence and Navier Stokes Equations. Proceedings 1975. Edited by R. Temam. IX, 194 pages. 1976.

Vol. 566: Empirical Distributions and Processes. Oberwolfach 1976. Proceedings. Edited by P. Gaenssler and P. Révész. VII, 146 pages. 1976.

Vol. 567: Séminaire Bourbaki vol. 1975/76. Exposés 471–488. IV, 303 pages. 1977.

Vol. 568: R. E. Gaines and J. L. Mawhin, Coincidence Degree, and Nonlinear Differential Equations. V, 262 pages. 1977.

Vol. 569: Cohomologie Etale SGA 4½. Séminaire de Géométrie Algébrique du Bois-Marie. Edité par P. Deligne. V, 312 pages. 1977.

Vol. 570: Differential Geometrical Methods in Mathematical Physics, Bonn 1975. Proceedings. Edited by K. Bleuler and A. Reetz. VIII, 576 pages. 1977.

Vol. 571: Constructive Theory of Functions of Several Variables, Oberwolfach 1976. Proceedings. Edited by W. Schempp and K. Zeller. VI. 290 pages. 1977

Vol. 572: Sparse Matrix Techniques, Copenhagen 1976. Edited by V. A. Barker. V, 184 pages. 1977.

Vol. 573: Group Theory, Canberra 1975. Proceedings. Edited by R. A. Bryce, J. Cossey and M. F. Newman. VII, 146 pages. 1977.

Vol. 574: J. Moldestad, Computations in Higher Types. IV, 203 pages. 1977.

Vol. 575: K-Theory and Operator Algebras, Athens, Georgia 1975. Edited by B. B. Morrel and I. M. Singer. VI, 191 pages. 1977.

Vol. 576: V. S. Varadarajan, Harmonic Analysis on Real Reductive Groups. VI, 521 pages. 1977.

Vol. 577: J. P. May, E_∞ Ring Spaces and E_∞ Ring Spectra. IV, 268 pages. 1977.

Vol. 578: Séminaire Pierre Lelong (Analyse) Année 1975/76. Edité par P. Lelong. VI, 327 pages. 1977.

Vol. 579: Combinatoire et Représentation du Groupe Symétrique, Strasbourg 1976. Proceedings 1976. Edité par D. Foata. IV, 339 pages. 1977.

Vol. 580: C. Castaing and M. Valadier, Convex Analysis and Measurable Multifunctions. VIII, 278 pages. 1977.

Vol. 581: Séminaire de Probabilités XI, Université de Strasbourg. Proceedings 1975/1976. Edité par C. Dellacherie, P. A. Meyer et M. Weil. VI, 574 pages. 1977.

Vol. 582: J. M. G. Fell, Induced Representations and Banach *-Algebraic Bundles. IV, 349 pages. 1977.

Vol. 583: W. Hirsch, C. C. Pugh and M. Shub, Invariant Manifolds. IV, 149 pages. 1977.

Vol. 584: C. Brezinski, Accélération de la Convergence en Analyse Numérique. IV, 313 pages. 1977.

Vol. 585: T. A. Springer, Invariant Theory. VI, 112 pages. 1977.

Vol. 586: Séminaire d'Algèbre Paul Dubreil, Paris 1975–1976 (29ème Année). Edited by M. P. Malliavin. VI, 188 pages. 1977.

Vol. 587: Non-Commutative Harmonic Analysis. Proceedings 1976. Edited by J. Carmona and M. Vergne. IV, 240 pages. 1977.

Vol. 588: P. Molino, Théorie des G-Structures: Le Problème d'Equivalence. VI, 163 pages. 1977.

Vol. 589: Cohomologie l-adique et Fonctions L. Séminaire de Géométrie Algébrique du Bois-Marie 1965–66, SGA 5. Edité par L. Illusie. XII, 484 pages. 1977.

Vol. 590: H. Matsumoto, Analyse Harmonique dans les Systèmes de Tits Bornologiques de Type Affine. IV, 219 pages. 1977.

Vol. 591: G. A. Anderson, Surgery with Coefficients. VIII, 157 pages. 1977.

Vol. 592: D. Voigt, Induzierte Darstellungen in der Theorie der endlichen, algebraischen Gruppen. V, 413 Seiten. 1977.

Vol. 593: K. Barbey and H. König, Abstract Analytic Function Theory and Hardy Algebras. VIII, 260 pages. 1977.

Vol. 594: Singular Perturbations and Boundary Layer Theory, Lyon 1976. Edited by C. M. Brauner, B. Gay, and J. Mathieu. VIII, 539 pages. 1977.

Vol. 595: W. Hazod, Stetige Faltungshalbgruppen von Wahrscheinlichkeitsmaßen und erzeugende Distributionen. XIII, 157 Seiten. 1977.

Vol. 596: K. Deimling, Ordinary Differential Equations in Banach Spaces. VI, 137 pages. 1977.

Vol. 597: Geometry and Topology, Rio de Janeiro, July 1976. Proceedings. Edited by J. Palis and M. do Carmo. VI, 866 pages. 1977.

Vol. 598: J. Hoffmann-Jørgensen, T. M. Liggett et J. Neveu, Ecole d'Eté de Probabilités de Saint-Flour VI – 1976. Edité par P.-L. Hennequin. XII, 447 pages. 1977.

Vol. 599: Complex Analysis, Kentucky 1976. Proceedings. Edited by J. D. Buckholtz and T. J. Suffridge. X, 159 pages. 1977.

Vol. 600: W. Stoll, Value Distribution on Parabolic Spaces. VIII, 216 pages. 1977.

Vol. 601: Modular Functions of one Variable V, Bonn 1976. Proceedings. Edited by J.-P. Serre and D. B. Zagier. VI, 294 pages. 1977.

Vol. 602: J. P. Brezin, Harmonic Analysis on Compact Solvmanifolds. VIII, 179 pages. 1977.

Vol. 603: B. Moishezon, Complex Surfaces and Connected Sums of Complex Projective Planes. IV, 234 pages. 1977.

Vol. 604: Banach Spaces of Analytic Functions, Kent, Ohio 1976. Proceedings. Edited by J. Baker, C. Cleaver and Joseph Diestel. VI, 141 pages. 1977.

Vol. 605: Sario et al., Classification Theory of Riemannian Manifolds. XX, 498 pages. 1977.

Vol. 606: Mathematical Aspects of Finite Element Methods. Proceedings 1975. Edited by I. Galligani and E. Magenes. VI, 362 pages. 1977.

Vol. 607: M. Métivier, Reelle und Vektorwertige Quasimartingale und die Theorie der Stochastischen Integration. X, 310 Seiten. 1977.

Vol. 608: Bigard et al., Groupes et Anneaux Réticulés. XIV, 334 pages. 1977.

Vol. 609: General Topology and Its Relations to Modern Analysis and Algebra IV. Proceedings 1976. Edited by J. Novák. XVIII, 225 pages. 1977.

Vol. 610: G. Jensen, Higher Order Contact of Submanifolds of Homogeneous Spaces. XII, 154 pages. 1977.

Vol. 611: M. Makkai and G. E. Reyes, First Order Categorical Logic. VIII, 301 pages. 1977.

Vol. 612: E. M. Kleinberg, Infinitary Combinatorics and the Axiom of Determinateness. VIII, 150 pages. 1977.

Vol. 613: E. Behrends et al., L^p-Structure in Real Banach Spaces. X, 108 pages. 1977.

Vol. 614: H. Yanagihara, Theory of Hopf Algebras Attached to Group Schemes. VIII, 308 pages. 1977.

Vol. 615: Turbulence Seminar, Proceedings 1976/77. Edited by P. Bernard and T. Ratiu. VI, 155 pages. 1977.

Vol. 616: Abelian Group Theory, 2nd New Mexico State University Conference, 1976. Proceedings. Edited by D. Arnold, R. Hunter and E. Walker. X, 423 pages. 1977.

Vol. 617: K. J. Devlin, The Axiom of Constructibility: A Guide for the Mathematician. VIII, 96 pages. 1977.

Vol. 618: I. I. Hirschman, Jr. and D. E. Hughes, Extreme Eigen Values of Toeplitz Operators. VI, 145 pages. 1977.

Vol. 619: Set Theory and Hierarchy Theory V, Bierutowice 1976. Edited by A. Lachlan, M. Srebrny, and A. Zarach. VIII, 358 pages. 1977.

Vol. 620: H. Popp, Moduli Theory and Classification Theory of Algebraic Varieties. VIII, 189 pages. 1977.

Vol. 621: Kauffman et al., The Deficiency Index Problem. VI, 112 pages. 1977.

Vol. 622: Combinatorial Mathematics V, Melbourne 1976. Proceedings. Edited by C. Little. VIII, 213 pages. 1977.

Vol. 623: I. Erdelyi and R. Lange, Spectral Decompositions on Banach Spaces. VIII, 122 pages. 1977.

Vol. 624: Y. Guivarc'h et al., Marches Aléatoires sur les Groupes de Lie. VIII, 292 pages. 1977.

Vol. 625: J. P. Alexander et al., Odd Order Group Actions and Witt Classification of Innerproducts. IV, 202 pages. 1977.

Vol. 626: Number Theory Day, New York 1976. Proceedings. Edited by M. B. Nathanson. VI, 241 pages. 1977.

Vol. 627: Modular Functions of One Variable VI, Bonn 1976. Proceedings. Edited by J.-P. Serre and D. B. Zagier. VI, 339 pages. 1977.

Vol. 628: H. J. Baues, Obstruction Theory on the Homotopy Classification of Maps. XII, 387 pages. 1977.

Vol. 629: W.A. Coppel, Dichotomies in Stability Theory. VI, 98 pages. 1978.

Vol. 630: Numerical Analysis, Proceedings, Biennial Conference, Dundee 1977. Edited by G. A. Watson. XII, 199 pages. 1978.

Vol. 631: Numerical Treatment of Differential Equations. Proceedings 1976. Edited by R. Bulirsch, R. D. Grigorieff, and J. Schröder. X, 219 pages. 1978.

Vol. 632: J.-F. Boutot, Schéma de Picard Local. X, 165 pages. 1978.

Vol. 633: N. R. Coleff and M. E. Herrera, Les Courants Résiduels Associés à une Forme Méromorphe. X, 211 pages. 1978.

Vol. 634: H. Kurke et al., Die Approximationseigenschaft lokaler Ringe IV, 204 Seiten. 1978.

Vol. 635: T. Y. Lam, Serre's Conjecture. XVI, 227 pages. 1978

Vol. 636: Journées de Statistique des Processus Stochastiques, Grenoble 1977, Proceedings. Edité par Didier Dacunha-Castelle et Bernard Van Cutsem. VII, 202 pages. 1978.

Vol. 637: W. B. Jurkat, Meromorphe Differentialgleichungen. VI, 194 Seiten. 1978.

Vol. 638: P. Shanahan, The Atiyah-Singer Index Theorem, An Introduction. V, 224 pages. 1978.

Vol. 639: N. Adasch et al., Topological Vector Spaces. V, 125 pages 1978.